Why Humans Matter More Than Ever

Why Humans Matter More Than Ever

MIT Sloan Management Review

The MIT Press
Cambridge, Massachusetts
London, England

This book was set in Stone Serif and Stone Sans by Jen Jackowitz. Printed and bound in the United States of America.

Library of Congress Cataloging-in-Publication Data
Names: MIT Sloan Management Review, issuer.
Title: Why humans matter more than ever / MIT Sloan Management Review.
Description: Cambridge, MA : MIT Press, [2019] | Series: The digital future of management | Includes bibliographical references and index.
Identifiers: LCCN 2018059568 | ISBN 9780262537575 (pbk. : alk. paper)
Subjects: LCSH: Human-machine systems. | Human-computer interaction. | Technology.
Classification: LCC TA167 .W49 2019 | DDC 620.8/2--dc23 LC record available at https://lccn.loc.gov/2018059568

10 9 8 7 6 5 4 3 2 1

Contents

Series Foreword

Books in the Digital Future of Management series draw from the print and web pages of *MIT Sloan Management Review* to deliver expert insights and sharply tuned advice on navigating the unprecedented challenges of the digital world. These books are essential reading for executives from the world's leading source of ideas on how technology is transforming the practice of management.

Paul Michelman
Editor in chief
MIT Sloan Management Review

Introduction: A Platform Greater Than Facebook

Paul Michelman

At *MIT Sloan Management Review*, we have published—and will continue to publish—volumes of content extolling the importance of digital transformation, with much of it focusing on the good that new technologies stand to deliver to both business and broader society. I, myself, am a techno-optimist. But there are also times when we need to step back, take stock, and seize just a bit more control over how our world is evolving.

Lately, many of us have been suffering a period of particular disquiet. The seemingly endless waves of technological and political disorder have been deeply unnerving, as disruptions in one arena feed turmoil in others.

How we choose to live our own lives is at stake as well. We have become public citizens almost by force at the same time that our trust in public institutions has plummeted. Our details are for sale—and we don't know to whom. We find ourselves looking for answers and action, for some sense of order to be brought to bear. Yet we are not certain from whom we expect this. It's all gotten very personal.

Now, take a breath and consider for a moment whether a vast swath of the world's population hasn't been experiencing an extreme version of this lack of agency their whole lives.

So let's agree to do something positive. There is a change necessary today that only humans can bring about, one in which we are not the forced reactors to technological advancement and political discord but the architects of our own platform.

Most of you reading this have influence. I encourage you to use it within your organizations and your communities. Lobby for sound, sustainable policy that creates broadly felt value. Demand that your companies look around the bend. Plenty of lip service has been paid to the need for leaders to stop managing for quarterly results. Let's choose now to act on that call. Dare to sacrifice a dividend for a development initiative, one that eyes the challenges of the years ahead, not just the weeks. Your shareholders are ready to hear your case.

Yes, there's a platform greater than Facebook, and there are ecosystems greater than Google. And we have not been tending to our biggest platform and most important ecosystems with appropriate care. Let's get back to taking the long view and embrace the huge challenge of harnessing technology to create a wealthier society, not just wealthier companies and individuals.

Now, more than ever, we need to look out for each other.

***

In this book, we bring together some of the best research and analysis from *MIT SMR* on how to move forward into the brave new digital world with nerve, effectiveness, and, most of all, humanity.

Making Technology Fit for Humans

In the first section, we explore how new technologies, including the most sophisticated types of artificial intelligence, depend on human collaboration if organizations are going to realize their full potential. Companies need to develop rules, principles, and clear ethical guidelines to structure the interactions that their smart objects have with humans.

We also need to think about how we regulate *ourselves* amid the noise of data and tidal waves of information. "Managing the Distraction–Focus Paradox" by Carsten Lund Pedersen makes this point: Those of us who hope to succeed as thinkers, managers, and innovators in a world filled with technology distractions must learn how to manage our most valuable personal resource—our attention.

How We Work

Next, we look at the big picture of how our work lives are being redefined by new technologies. We need to understand the ways that our jobs are evolving—and the factors behind those changes—and we need to embrace the need to adapt and become more collaborative.

Leaders must fully seize their central role in preparing their organizations for the coming world of work. They need, as Lynda Gratton argues in "Face the Future of Work," to be "deeply aware—right now, not down the line—of the transition taking place." Leaders must actively engage by acknowledging to their teams that work is changing, by taking responsibility for helping employees learn new skills, and by role modeling flexibility around alternate ways of work such as job sharing. For many leaders, these will be difficult challenges to face.

How We Manage

In the final section, we delve into how technology is changing our work lives in the day to day. Virtual teams and collaborations that take place across professions, geographies, and industries are, of course, all made easier by technology. But the skills needed to capture the full value of these multifaceted collaborations don't come naturally to many people.

"We don't always understand one another's expertise or even one another's outlook," Harvard Business School's Amy Edmondson notes in "The Leadership Demands of Extreme Teaming." Empathy and curiosity will take leaders only so far, she maintains: "Leaders must also have a high level of self-awareness to keep reminding themselves of the things that they are missing." Each of us thinks that we see is "reality," when in fact we don't know everything.

***

Edmondson is absolutely right when she says that we can all learn to be curious, empathic, and interested in other people's perspectives, but she's also right when she says that this kind of humility is not a given. Ironically, as we move forward to manage ever more complicated systems and situations, we may find that it pays to acknowledge the need to learn as we go. Saying "I don't have the answer" may not come easily, but it may be the best way to get to a place where you do.

I

**Making Technology Fit
for Humans**

1

Humanizing Tech May Be the New Competitive Advantage

Bala Iyer, Kristen Getchell, and Fritz Fleischmann

At Google's annual developer conference, Google I/O 2018, the company's CEO, Sundar Pichai, proudly demonstrated Google Duplex, a new artificial intelligence voice technology, making a remarkably human-sounding reservation over the phone. The problem was that the actual human on the other line did not know she was interacting with a bot. Only after Google faced backlash over concerns about this kind of deception did the company agree to release Duplex with disclosure built in.

As software continues to "eat the world," the potential for privacy and ethics violations increases. It's clear that technology executives and managers need to recognize the industry-wide factors that have contributed to the current fractured state of customer trust and move toward a framework that puts users first.

First, let's examine some of the contributing factors of the current status quo.

Believing Moore's law for too long Intel cofounder Gordon Moore famously predicted that computing power, measured by quantity of transistors, would double every year, leading to exponential growth in this field. Moore's law persisted throughout the hardware and software age, and only recently have we begun to consider its demise. With such a focus on growth and velocity of innovation, many technologists have found themselves ill prepared to consider the impact of their technology.

Favoring the individual company over the collective users In the tragedy of the commons, individual rationality and collective rationality are at odds with one another and are contradictory. This same conundrum exists today in tech—companies capture and use customers' personal information but fail to show concern about the overall damage they cause by their individual actions.

Companies have acted in favor of increasing market share, but in the process have eroded the confidence and trust of customers. This was quite clear as we watched Facebook's Mark Zuckerberg grilled by Congress over the consequences of Cambridge Analytica's data privacy scandal with Facebook user data.

Leading with tech first, questions second (or not at all) The rise in artificial and augmented intelligence has led to a proliferation of technologies that create, mimic, and facilitate conversation. This means designers are now introducing empathy, personality, and creativity to machine-human interaction in ways that affect user experience. The relationship a machine has with (and to) a user becomes a new competitive advantage.

Everyday objects are now becoming smart objects with the ability to interact with humans. What are the guidelines for structuring these conversations? Google has raised the question of whether users should be informed that they are interacting with a computer. What ethical rules should be in play when it comes to using these products, whether it's a voice assistant, a TV, or even a car?

Companies that excel in addressing these questions to gain the trust of users will be given the opportunity to offer new products and services to those users. The key ingredient here—and this cannot be stated too often—is trust.

Moving from a "Can We?" to a "Should We?" Framework

Technology and business experts must do a better job of anticipating challenges before making decisions, by asking key user-centered questions before launching new products into the market. The following questions on a technology's impact must be systematically addressed before bringing it to market:

- Will this technology result in overall good?
- What might be some unintended consequences of this technology?
- What are the social and ethical impacts of the technology?
- Will this technology augment human intellect, disrupt it, or substitute for it?
- How could this technology be used negatively against users?

Technologists won't be able to answer these questions by themselves—which brings us to the most important question all executives need to ask: What leadership structures do we need to have in place to guide the future evolution of the technology while controlling for unintended negative consequences?

We argue that the answer to this last question needs to be more than simply "we need more engineers." Instead, it is important for leaders to embrace the following six principles and ensure they are introduced at every level of the organization.

1. **Assume responsibility.** Companies need to assume ethical and legal responsibility for the impact of their technology on society. The burden of proof should be on companies to provide reasonable assurances that they have scrutinized the impact that their products would have.

2. **Offer transparency.** It is important that individuals have the ability to access information about any technology they use. Companies should provide frequent impact disclosures on all developing technology, including answers to the questions about their impact. Companies working on the cutting edge of AI should be subject to external review.

3. **Give users the right to be forgotten.** If customers would like to leave a product or system, they should be able to do so easily, with one click. This would apply to user accounts or personal and transactional data stored by a company. With the European Union's General Data Protection Regulation having taken effect in May 2018, this is now a legal requirement for companies doing business in Europe, not an option.

4. **Anticipate technology adoption challenges.** Questions about a technology's impact should not be addressed only after the technology has been developed or in the case of public backlash. Concerns of intended and unintended impact need to be addressed during the engineering process and embedded in the development of a technology. Ethical considerations can no longer be an afterthought.

5. **Conduct experiments.** Companies must seek empirical evidence to determine how people react to new technology or

changes in existing technology. When introducing technology-enabled product features, companies should conduct statistical experiments to determine if users like the changes. For example, if Facebook decides to provide automatic updates on news feeds, it must first conduct multiple tests with a subset of users and then release those data to the public.

6. **Assemble a team of diverse thinkers.** Tech firms must integrate individuals with expertise outside of business and technology into decision-making points across organizations. New skill sets are required when, for example, companies trying to develop conversational commerce technologies seek to design a user experience that is more accessible and humane. Linguists, scriptwriters, human development specialists, sociologists, physicians, scientists, psychologists, and ethicists can help to evaluate the quality of interaction and appropriateness of responses, how machines make users feel, and how technology could impact society. Technology projects power, and how that power should be used is not a technological but an ethical, social, and political question.

In summary, it's time to stop thinking of Moore's law as if it were a natural law. Humanizing technology should be a core capability of companies for both ethical and competitive reasons. By striking a balance between technological innovation and concerns for users, organizations can achieve a new competitive advantage—one that legacy companies may, in fact, be better poised to gain as many digital natives face rebuilding customer trust as their next challenge.

2

Managing the Distraction–Focus Paradox

Carsten Lund Pedersen

In the time you've set aside to read this article, you're likely to check your phone. You'll probably see notifications for emails or text messages pop up on your lock screen. You won't resist. Once you've started thumbing through your apps, you'll check Twitter, too. If you use Twitter as your media feed, you may click through to an article about blockchain or vacations in Barbados. I'll be lucky if you make it back here.

Nicholas Carr, author of *The Shallows: What the Internet Is Doing to Our Brains*, would have you believe that your behavior is a serious problem, that the ephemera of the internet are hijacking your ability to concentrate and think.[1] I disagree—or rather, I'd argue that, in today's workplace, the seductive clamor of the web is a reality from which there's no retreat. In the age of big data and ever more powerful processors, we must absorb more data at faster speeds. Those who'll succeed in this distraction-filled world as thinkers, managers, and innovators will need to combine two seemingly opposing traits. They must be able to absorb diverse information from a wealth of sources, and they must be able to focus intensely. I call this the distraction–focus paradox. While these two qualities seem contradictory, together they make up the skill set for managing your most valuable personal resource—your attention—in a hyper-connected age.

Yes, these abilities have always been important—but their combination will become more so in the coming years, as social media and mobile computing continue to advance. (See "Skill Set for a Connected World," which presents the net effect of differing combinations of these essential skills.)

Knowledge workers need diverse information. Research has repeatedly shown that diversity in mental models—that is, how you interpret and see problems—leads to better problem solving and more innovation.[2] That's a theme that courses through *Misbehaving: The Making of Behavioral Economics*, the memoir of Richard Thaler, the Charles R. Walgreen Distinguished Service Professor of Behavioral Science and Economics at the University of Chicago Booth School of Business and the 2017 winner of the Nobel Prize in economics.[3] As a young scholar, Thaler kept noticing anomalies that defied standard economic models, like

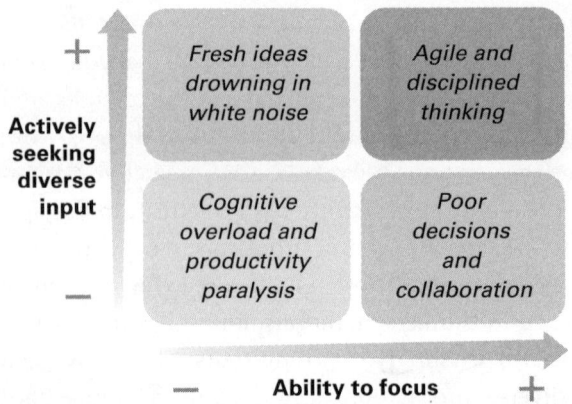

Skill Set for a Connected World
"Productive distraction" balances curiosity and concentration.

the so-called endowment effect—the tendency of people to over-value things they already own. Even as an established scholar, his curiosity has ranged widely, as he has published papers on such topics as why NFL teams make irrational decisions in the annual player draft. To help make sense of such phenomena, he collaborated with psychologists.

People like Thaler who seek out varied inputs have been shown to be consistently better forecasters than those who rely on more limited information diets.[4] And the need to avail one-self of a variety of perspectives has only increased. One of the dangers of the rise of social media is that people's networks are insufficiently diverse—and consequently risk becoming echo chambers.[5] This trend is exemplified by the notions of filter bub-bles and "fake news."[6] We connect more and more, but often only with people or publications that share our views.

Compare the experience of browsing at a bookstore a decade or so ago with buying a book online today. In an old-fashioned store, as you ambled over to the business section, you might happen across the archaeology and anthropology books. If you had even a glimmer of interest, you'd find yourself studying the spines. Maybe you'd end up buying Jared Diamond's surprise best seller, *Guns, Germs, and Steel*.[7] Had you read that 1997 tour de force, you would have learned about the original domestica-tion of plants and animals and the evolution of disease immu-nities and how both of those influenced the distribution of the world's wealth. Today, if you search for the latest business best seller on Amazon, you're highly unlikely to receive such an eso-teric recommendation.

Tapping into diverse networks also fosters innovation. Much innovation has originated from individuals relying on collabo-ration with open networks—and research has even shown that people with more diverse Twitter feeds tend to generate better ideas.[8] So people need to train themselves to seek out sources with heterogeneous views. Indeed, when I refer to "distraction," you could think of that partly as the cognitive load that comes from immersing yourself in a more diverse network.

And yet, it's also important to be able to focus intensely on a specific problem, particularly as expectations of instant responses to emails, alerts, and notifications nag at our attention. Ours is the age of distraction, good and bad. The web blesses us with news from Belarus and the latest advances in biology and bedevils us with listicles and personality quizzes on such weighty topics as which dog breed or which character from *The Simpsons* you most resemble.

As the digital sirens continue to sing, maintaining energized, deep focus matters even more. Some of the proponents of this line of thought, including Cal Newport, author of *Deep Work*, have argued that the ability to focus on a demanding task is the way to differentiate yourself in a distracted world.[9] This kind of focus entails winnowing the demands and "productive distractions" vying for your attention and time. It also requires the ability to shift between perspectives: seeing the details and the broader context. If you can focus in this way, you can prioritize *what* to think about (you can better plan) and you can know *how* to think about it (you can better process). But being focused does not mean behaving like a robot. Focus is the deliberate deployment of your attention. You lock in, rather than zone out.

As is so often true, too much of either of these information-age virtues isn't beneficial, either. If you ramble around the web, pointing and clicking willy-nilly without a goal or guidelines—without a focus—the white noise will block out your ability to hear anything worthwhile. You won't devote enough time to critical tasks, nor will you distinguish important issues from irrelevancies. But, if you are too focused and deprive yourself of varied views, you run the risk of lacking creativity and insight. Research has shown that excessive focus can exhaust a person's attention and lead to ill-conceived decisions and less collaboration.[10]

Yet having too little information and too little focus seems even worse. Who'd settle for that? Of course, that's the situation we so often encounter in our digitized, socially connected world: We're bombarded with tweets, emails, and Facebook and LinkedIn requests from friends and colleagues, preventing us from finding time to seek out fresh insights or to focus fully on the tasks we consider most important.

The goal is, of course, the Golden Mean—a balance between diversity of input and intensity of focus. If you can achieve that, you're better equipped for our distracted age. This skill set can be understood as a form of meta-cognition—like having a personal project manager inside your head. These skills are analogous to qualities possessed by the best leaders and organizations: consistency *and* agility.[11]

People who can balance curiosity and concentration fit into the metaphor of the "T-shaped professionals" popularized by Ideo, the design consultancy. The vertical leg of the T conveys focus (expertise and insight) while the horizontal one conveys open-mindedness (empathy and collaborative curiosity). According to Ideo's CEO, Tim Brown, T-shaped people can focus deeply on their particular domains while also interacting productively with colleagues from different disciplines. These complementary characteristics are often needed at Ideo when people solve specific problems.[12]

So how can you enhance your T-shaped qualities? First, you need to assess your abilities and position yourself in the skill set matrix. Knowing your strengths and weaknesses, you can then seek to improve. If you score low on seeking diverse input, challenge yourself to find new sources of information that broaden your knowledge and contradict your assumptions. Or try being your own devil's advocate and asking yourself, routinely, what would be the opposite perspective on this problem—and what type of information would support it?

If you score low on focus, turn off your phone (or at least your notifications) and carve out blocks of time for undisturbed thinking and reading. Philanthropist Bill Gates used to make this a practice when he was running Microsoft. He'd take a "think week" twice a year, retreat to a lakeside cabin, read, and ponder his company's future.[13] These days, Gates posts thoughts about the books he has read lately on his blog, GatesNotes.com, and invites favorite authors to his office for lunch.[14] And Gates' good friend, billionaire investor Warren Buffett, is likewise famed for being a "learning machine" who, by his own admission, often sits in his office and reads all day.[15]

You can also team up with collaborators who have strengths that complement yours. Or you can just work on getting better: Like many skills, self-questioning and focus can be improved through deliberate practice.[16]

We're living in an age of uncertainty, driven by technological and social change: Cars are driving themselves, drones will soon be delivering packages, and the "free-agent economy" is demanding professionals who can reinvent themselves throughout their careers. To thrive in these turbulent times, you must be capable of "distracted focus."

3

Want the Best Results from AI? Ask a Human

Bhaskar Ghosh, Kishore Durg, Arati Deo, and Mallika Fernandes

Companies of all kinds are adopting artificial intelligence and machine-learning systems at an accelerated pace. International Data Corp. (IDC) projects that shipments of AI software will grow by 50% per year and will reach $57.6 billion in 2021—up from $12 billion in 2017 and just $8 billion in 2016. AI is being applied to a range of tasks, including rating mortgage applications, spotting signs of trouble on power lines, and helping drivers navigate using location data from smartphones.

But companies are learning the hard way that developing and deploying AI and machine-learning systems is not like implementing a standard software program. What makes these programs so powerful—their ability to "learn" on their own—also makes them unpredictable and capable of errors that can harm the business.

AI's Challenge: It's Susceptible to Learned Bias

We frequently hear stories of AI gone awry. For instance, lenders are grappling with AI systems that unintentionally "learn" to deny credit to residents of certain ZIP codes, which is a violation of bank "redlining" regulations. Or consider an online translation program that, when asked to translate the phrase "She is a doctor, and he is a nanny" into Turkish and then translate it back to English, spits out: "He is a doctor, and she is a nanny."

These bias-induced situations can have serious business consequences. When AI was being used in back-office applications, the chance of bias creeping in was limited, and so was the potential damage. Now AI is being used extensively both in management decision support and customer-facing applications. Companies risk damaging people's reputations and lives, making strategic wrong turns, offending customers, and losing sales. And the cost of AI mistakes—whether they come from bias or flat-out error based on unreliable data or faulty algorithms—is rising.

The lesson here is that AI systems, for all their amazing powers, still need continuous human intervention to stay out of trouble and do their best work. Indeed, companies are finding that they get the most out of investments in AI and other automation tools when they think in terms of humans and machines working together, rather than dividing work between humans and machines and letting the machines operate independently.

When conventional software is installed, procedures and rules are set in stone by human developers. By contrast, an AI system develops its own rules from patterns in the data it is crunching. And, as some companies have learned through real-life situations, AI systems can jump to the wrong conclusions.

Three Guiding Principles for Successfully Adopting AI

Therefore, before diving into AI systems, companies should consider three principles that can greatly improve the chances for a successful outcome:

Principle 1: Remember, humans and machines are in this together Nowhere is human–machine collaboration more relevant than in installing and maintaining AI systems. Human assistance is needed to teach and monitor AI systems properly and keep them from drifting into dangerous territory over time. This is not a job for IT departments alone—it requires both technical expertise and business understanding.

Training and monitoring the ongoing performance of an AI system requires that employees who are experts in software collaborate with colleagues who rely on AI systems to do their work. As users of AI output, these colleagues are in a position to spot changes in how the program is performing and can act on any issues that arise. Similarly, while a road-mapping application might use AI to plot efficient driving routes, a human driver can override the system's choices based on knowledge of rush hour patterns or road construction.

Principle 2: Teach with (a lot of) data AI systems learn by finding patterns in training data through various algorithms. Typically, this is done with historical data and involves experimenting with different models. The trained models are statistically evaluated, and the best-performing model is selected to be deployed into production.

This means that AI has a lot to learn. For example, a business often needs to evaluate how the brand is doing on quality and service based on unstructured data like comments on Twitter, news stories, Facebook posts, online reviews, and the like. The model must be trained with real-time data and programmers (or ordinary employees who have learned how to train AI systems) and taught rules that the program would not pick up on its own. Programmers would have to teach the system how to understand the true meaning and validity of consumer comments.

Machines, for instance, don't understand sarcasm (although Israeli scientists say they have developed a program to identify sarcastic comments). Other challenges in accurately parsing user-generated content include interpreting specific words differently depending on the context—"hot" would have a positive connotation in the context of food but could have a negative sentiment in the context of how comfortable the restaurant was.

Training can be labor-intensive up front, but with a well-structured methodology for developing unbiased training data, training time of the AI system can be reduced by 50%, according to Accenture's internal research.

Principle 3: Continually test the results   With AI programs, testing not only is critical prior to release, but also becomes an ongoing routine. Managers need to be confident that the system will deliver accurate results from a variety of data.

Traditional software testing is determinate—you need to test only a finite number of scenarios. Once the program has been tested for all possible scenarios, it is guaranteed to work. But with AI and machine learning, you can't predict every scenario. You must continually monitor and test the system to catch data biases as well as biases that develop in the algorithms that the programs use to make judgments.

You can test for data bias by using more than one set of data—for example, a loan-application system based on historical data will only perpetuate the biases inherent in that data, which will likely show that members of certain groups in the population have not qualified for loans. To correct for this bias, the system must be tested and retrained with additional data. For instance, to make sure that the algorithm that monitors consumer sentiment about your brand is working properly, you can test the same set of data with different algorithms.

The New Normal: Teaching, Testing, and Working with End Users

Effectively deploying AI requires a new conception of how software is developed, installed, and maintained. Teaching, testing, and working with end users of AI output must become a way of life, enabling AI systems to continually operate more responsibly, accurately, and transparently—and allowing businesses to create collaborative and powerful new members of the workforce.

II

How We Work

4

Four Ways Jobs Will Respond to Automation

Scott Latham and Beth Humberd

There is no question that automation is changing the nature of work. But are the robots really coming for your job?

One of the most popular narratives is that low-paying jobs are doomed, while college-educated professions will remain largely untouched. Analysts often focus on wages and education as the primary predictors of job evolution, along with organizations' potential to increase efficiency and reduce costs by changing or cutting jobs. But our research points to a more nuanced explanation.

A review of the academic literature and public discourse on automation revealed limited consideration of risks by profession. So we did our own comparison, coding 50 professions (including many from our literature survey) according to the type of value jobholders delivered and the skills they used to deliver it, to create a framework that helps workers assess what kind of threat automation poses for them. We identified four paths of evolution—jobs will be disrupted, displaced, deconstructed, or durable—and found that value is more predictive of change than wages, education, efficiency, cost, or other factors.

Counter to popular belief, it's not necessarily blue-collar or non-college-educated workers who will be most threatened by automation in the coming decades. Our analysis suggests that a plumber may see less disruption than a legal professional. Simply instructing everyone to engage in continuous education and skill development is remiss. Workers must understand the four paths of job evolution—and the factors behind each path—if they hope to adapt.

Understanding the Four Paths

A jobholder uses a core set of skills to deliver value in some form to a recipient—either externally to a customer or within an organization. Jobs evolve as those consumers' perceptions of value fluctuate along two dimensions: core skills and delivery mechanism, or what we call "value form."

For some jobs, core skill sets include a specific knowledge base or craft. Others involve people skills and the ability to build relationships rather than technical expertise. Skills that can easily be standardized, codified, or routinized are most likely to be automated. Those that involve hands-on or real-time problem solving are less so, because developing tools sophisticated enough to handle such ambiguity is either too cost- and labor-intensive or technologically out of reach. For example, while an electrician's skills may seem vulnerable to automation, the application of those skills varies widely according to the unique circumstances of every client. This degree of customization would be difficult to automate.

A skill set provides value only when it is delivered to a recipient, however, and the delivery mechanism may be transformed. Here's an example: A professor's core skill set is expertise in a certain domain. Such expertise has traditionally been delivered to consumers (students) through in-person classes. However, online platforms and massive open online courses, or MOOCs, offer new vehicles through which learning can occur. The core skill remains the same, but technology is shifting the value form as adaptive software and virtual tutors offer highly personalized instruction and support to growing numbers of students with diverse needs. And computer-directed learning will continue to improve with the increasing sophistication of automation and AI.

We identified the four ways automation will affect jobs by separately assessing the degree of threat to each profession's core

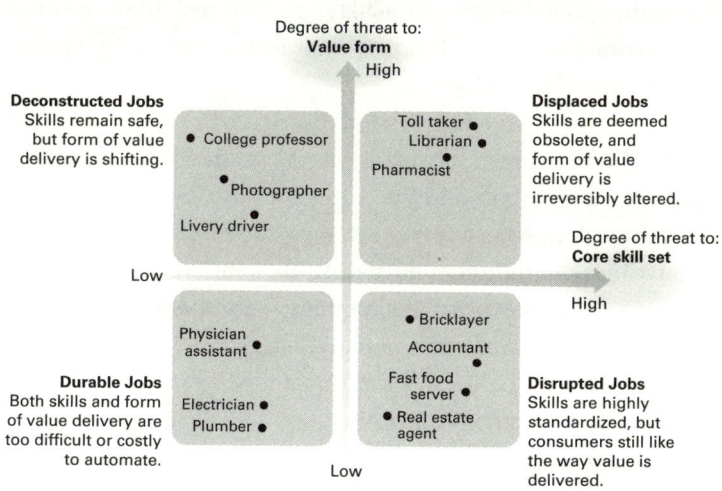

Which Professions Are Most Vulnerable to Automation?

Threats should be assessed along two dimensions: How replaceable are the core skill sets? And how much of a shift is there in the way value is delivered?

skill set and value form. In "Which Professions Are Most Vulnerable to Automation?" we describe those paths to evolution and suggest strategies for navigating each one.

Disruption   Disruption occurs when the skills in a job are highly standardized yet the consumer prefers to receive value in the same form. It typically follows a reduction in the production costs of goods or services due to increased efficiency. For example, fast food workers' core skill sets are highly threatened by the implementation of self-ordering stations and apps where customers place their own orders. Food preparation in this setting is also highly standardized and may eventually be automated as well, disrupting workers in both checkout stations and kitchens. Although these workers' skills are threatened, the consumer will continue to receive the same value form—fast food prepared consistently and quickly.

Some highly skilled professionals, such as real estate agents and legal professionals, are experiencing similar disruption from house-hawking robots and the automation of document reviews and other routine legal tasks (although the more nuanced work of advising clients and negotiating in court requires human lawyers, at least for now). Accountants—another example—are seeing the automation of company ledgers and other types of financial data. Value form is not threatened because consumers still need access to their financials, but the skills used to generate those financials are vulnerable.

Finding transitional roles in which human involvement remains necessary is one adaptive solution. As large-scale automation continues to spread, consumers will have to learn to interact with nonhuman providers and adopt new routines. Disrupted workers can function as a bridge, ensuring that value is delivered to end users in its current form as processes are automated. For example, bricklaying robots are much faster and possess more stamina than their human counterparts. But for now, human bricklayers are necessary to complement and safeguard the robots' abilities, read blueprints, and do corners.

Displacement With displacement, the core skills of a job are deemed obsolete and the value form is irreversibly altered. Toll takers and telephone operators have already experienced displacement, but even highly skilled professions are not immune. Take pharmacists. They fill prescriptions, deliver them to consumers, and answer questions at brick-and-mortar pharmacies. Yet as more prescriptions are filled online and delivered through the mail, the value form and core skills of human pharmacists are increasingly fulfilled by automated processes. Other jobs facing displacement include librarians (for similar reasons) and software developers (because the skill of writing code is easily standardized, and thus value form has shifted away from in-house development to open platforms such as the cloud).

Retraining is often recommended for displaced workers, but that doesn't always mean more formal education. They should focus on quickly acquiring the most relevant skills in an area with a relatively stable value form. In a volatile job market, lengthy programs that require years to complete (such as extra bachelor's degrees) are likely not the best approach. Micro-credentialing programs—competency-based certifications, mini-degrees, and digital badges—deliver qualifications more quickly and offer more options on the path to a degree along with a sense of accomplishment as individuals obtain marketable skills fast. We suggest targeting high-growth sectors that need workers. A timely example is cybersecurity—a rapidly growing field where trained workers (who can qualify through certificate programs) are in demand.

Deconstruction   In the case of deconstruction, the core skill set remains safe, but the value form is threatened. Take, for example, taxi or limo drivers, or anyone who operates a car service. Livery drivers' skills are central to the value delivered to customers— getting from point A to point B safely and efficiently. While those skills may be threatened by driverless automobiles at some point, human drivers will likely be a necessity in the near term. Yet the value form has already shifted. Traditionally, the value of livery transportation was offered as part of a centralized fleet— drivers were employed by a handful of taxi management companies within a city. Now, the same value is being delivered by Uber, Lyft, and others in the decentralized sharing economy. Photographers and professors are facing similar deconstruction. Their skills remain important, but consumer delivery preferences are changing.

When facing deconstruction, adapt your skills to new value forms. While this sounds easy enough, the biggest impediment is resistance to change. It is well documented, for example, that many faculty resist online education as a new model for sharing knowledge and expertise with students. Livery drivers would be wise to adjust to evolving transportation norms instead of following these professors' lead. When a new value form becomes central to consumers' expectations, you have a choice: acclimate or fade into obsolescence.

Durability   Often lost in workforce analyses is the fact that many jobs will remain unchanged for the foreseeable future, including some lower-wage jobs. We refer to jobs as durable when neither the core skill set nor the value form is under significant threat. Electricians and plumbers are highly durable professions because the work is rarely routine and the cost to develop a technology that could deliver value in the same form—hands-on problem solving—is excessive. Another example is the physician assistant. The skills associated with this job—medical training, insurance industry insight, bedside manner—will likely become more important as broader technological advancements require fewer doctors to treat more patients. Doing much the same work for less money, physician assistants may just disrupt the role of doctors.

The key for people in durable jobs is to avoid complacency by keeping an eye on tomorrow. Consider whether consumers' future preferences are more likely to threaten your profession's core skill set or its value form. Be aware that any job (including those discussed here) could drift from one evolution path to another over time. Thus, the framework we've described is a tool to be consulted regularly, even if your job is durable now.

It's difficult to tell which jobs will be disrupted, displaced, deconstructed, or durable further down the road, but we believe that the basic framework presented here will hold up to changing times. While others have acknowledged that automation will affect jobs in different ways, our focus on jobs as a function of value creation offers an explanation of the underlying dimensions at play. Understanding core skills and value form as the key units of analysis will help jobholders of all types respond to workforce changes currently underway—and tackle those that are impossible to predict.

5

How AI Can Amplify Human Competencies

Ken Goldberg, interviewed by Frieda Klotz

Though artificial intelligence systems are already becoming a part of daily life, recent debates about AI and the future of work have gained a sense of urgency. The late Stephen Hawking worried that humans "couldn't compete, and would be superseded" by machines, while Tesla founder Elon Musk has suggested that competition in AI could lead to World War III. *The Economist* reported earlier this year that nearly half of the jobs in 32 developed countries surveyed by the Organisation for Economic Co-operation and Development (OECD) were vulnerable to automation, declaring "a wave of automation anxiety has hit the West."

Ken Goldberg, professor and department chair of industrial engineering and operations research at the University of California, Berkeley, is pushing back on all of that. Instead of embracing the notion that robots will surpass humans and replace us in the workforce (a concept referred to as "singularity"), he argues for "multiplicity"—a hybrid view of how new technologies and people might work in partnership toward human goals. To an extent, he says, this is how AI is already starting to function.

MIT Sloan Management Review correspondent Frieda Klotz spoke with Goldberg about a future in which AI is a complement, not a threat, to workers. What follows is an edited and condensed version of their conversation.

MIT Sloan Management Review: What areas of robotic technology is your lab currently working on?

Ken Goldberg: We're developing robot software for tasks as wide-ranging as warehouse order fulfillment, home decluttering, and robot-assisted surgery. What's common to all the work we're doing is the idea of algorithms and learning for robots, improving our ability to analyze data and examples and then use that to build control policies—or models—for how robots can move.

The area I've been working on for 35 years is robot grasping—how to reliably pick up objects. It's easy for humans, but it's a problem for robots. Basically, every robot is still a klutz, and that's a big challenge if you want to develop one that will declutter a home or pack boxes in a warehouse.

Can you talk about your concept of multiplicity?

People keep saying we're on the verge of a transition, the singularity, when computers will take over. There's a sense that AI is a magical technology that's going to transform industries and replace humans, putting people out of work. But we're not anywhere near that point.

There are really good technologies and many interesting developments, and in some domains machines can be better than humans. Machines are very good at precision; they're very good at calculating numbers and pattern recognition. But there are several domains in which machines, and especially robots, don't excel. The most advanced robotic grasping technique isn't as deft as a 3-year-old! I'm concerned that people have expectations that are out of line with the current reality—and that these will distract us from what we should be worrying about and planning for. That's what led me to multiplicity, the idea that we'll see new partnerships between teams of humans and machines. Most of the systems that we use actually arise from human interaction. And this is already happening every day—for example, when by clicking on results, we give Google's search algorithm feedback that it then uses to refine future results.

Multiplicity requires diversity. If you look at a body of thinking called ensemble theory, you can prove that diversity is helpful for a machine-learning system. The relationship is something you can formulate mathematically. That's really exciting, because it's consistent with what we're starting to find about groups of humans: that if you have a diverse group of people, you get better, more creative ideas, more insights, and better outcomes.

We'll see different kinds of diversity, then – not just between people, but with people and robots putting their efforts together.

Exactly. Qualities like intuition, empathy, creativity are all very human—we're very good at looking at holistic situations, generalizations—and we can blend those qualities with the precision that machines provide.

We should be celebrating this, because it literally leads to better decisions and better processes.

In the next few years, how might robotics not be as useful as people expect?

People claim that we're going have autonomous trucks, which would eliminate truck driver jobs. They say this about Uber drivers or Lyft drivers too, but this is not going to come to pass.

We will make some progress; you can drive for good stretches on the freeway today with a robotic system. But there are so many complexities about driving in a city or a suburban environment that make it so much harder, especially if you're in a truck, because there are narrow and winding streets to navigate. We're going to need human truck drivers for the foreseeable future—for the rest of my lifetime and my kids' lifetimes.

Another example is that some claim there's no future for journalists. Computer systems take data about sporting events and then generate stories, which read reasonably well. That's because they can identify patterns and put numbers and results into those patterns, and it may work to an extent. But machines don't have the ability to pick up what is really interesting about a sporting event, the particular nuances of what's going on, or make analogies about what the teams are doing.

Aren't machine-learning teams working on these kinds of distinctions?

They are, but realistically they are years away from making it happen. What robots are great at are jobs that no one else wants to do—the dirty, dull, and dangerous jobs. I do think we'll have our decluttering robot that can tidy up around our homes in the next 10 years, at a price we can afford. Robots will also excel at tasks like washing windows on skyscrapers.

When it comes to more specialized fields like medicine, some of my work uses data from human surgeons and inferred models to develop robots that can perform suturing or remove fragments—tasks considered tedious by most surgeons. This gives physicians the ability to be focused and present and have more attention for the things that matter most.

What could business leaders be doing to allow these sorts of partnerships to flourish in their organizations?

CEOs should appreciate the value of the people who work for them and reassure employees that AI systems can actually help them do their jobs *better*, instead of replacing them.

AI will be able to perform many of the duller office tasks. Think of the pain points that hinder workers from getting on with the more important parts of their jobs—scheduling meetings, transcribing, taking notes, summarizing and indexing documents. What CEOs should be thinking about is how these tools can enhance the performance of employees.

Is there any risk that you are underestimating machines and their abilities?

I could be wrong, of course. But I have not seen any evidence that a computer is capable of innovation and creativity. Robots can be programmed to behave in a way that mimics human inventiveness, but they're unable to innovate spontaneously, to exchange ideas the way people do, to forge truly new insights or designs, and to recognize them as such. Doing this requires a vast understanding of what is normal and what isn't, which we don't know how to formalize.

It's one element of the Turing test, which examines whether a machine can keep up its end of an interesting conversation in a way that's indistinguishable from human intelligence. We're not even close; by that measure, we don't have intelligent machines, and we haven't made any progress, really, in 60 years. All the developments in AI are exciting, but that human-level frontier is still as hard to breach as it was decades ago.

Why do you think people have latched onto the idea of singularity when it may not accurately represent technological advances?

Even in the beginning of the 20th century, when automation came out, there was talk about robots taking over. It's cyclical.

People say this time is different—the technology is different. Yes and no. The fact is, we do have faster computers, we have a lot more data to work with, and we have made some progress. But in the most important ways, machines are nowhere near surpassing humans.

6

How Human–Computer "Superminds" Are Redefining the Future of Work

Thomas W. Malone

The ongoing, and sometimes loud, debate about how many and what kinds of jobs smart machines will leave for humans to do in the future is missing a salient point: Just as the automation of human work in the past allowed people and machines to do many things that couldn't be done before, groups of people and computers working together will be able to do many things in the future that neither can do alone now.

To think about how this will happen, it's useful to contemplate an obvious but not widely appreciated fact. Virtually all human achievements—from developing written language to making a turkey sandwich—require the work of groups of people, not just lone individuals. Even the breakthroughs of individual geniuses like Albert Einstein aren't conjured out of thin air; they are erected on vast amounts of prior work by others.

The human groups that accomplish all these things can be described as *superminds*. I define a supermind as a group of individuals acting together in ways that seem intelligent.

Superminds take many forms. They include the hierarchies in most businesses and other organizations; the markets that help create and exchange many kinds of goods and services; the communities that use norms and reputations to guide behavior in many professional, social, and geographical groups; and the democracies that are common in governments and some other organizations.

All superminds have a kind of collective intelligence, an ability to do things that the individuals in the groups couldn't have done alone. What's new is that machines can increasingly participate in the intellectual, as well as the physical, activities of these groups. That means we will be able to combine people and machines to create superminds that are smarter than any groups or individuals our planet has ever known.

To do that, we need to understand how people and computers can work together more effectively on tasks that require intelligence. And for that, we need to define intelligence.

What Is Intelligence?

The concept of intelligence is notoriously slippery, and different people have defined it in different ways. For our purposes, let's say that intelligence involves the ability to achieve goals. And since we don't always know what goals an individual or group is trying to achieve, let's say that whether an entity "seems" intelligent depends on what goals an observer attributes to it.

Based on these assumptions, we can define two kinds of intelligence. The first is specialized intelligence, which is the ability to achieve specific goals effectively in a given environment. This means that an intelligent entity will do whatever is most likely to help it achieve its goals, based on everything it knows. Stated even more simply, specialized intelligence is "effectiveness" at achieving specific goals. In this sense, then, specialized collective intelligence is "group effectiveness," and a supermind is an effective group.

The second kind of intelligence is more broadly useful and often more interesting. It is general intelligence, which is the ability to achieve a wide range of different goals effectively in different environments. This means that an intelligent actor needs not only to be good at a specific kind of task but also to be good at learning how to do a wide range of tasks. In short, this definition of intelligence means roughly the same thing as "versatility" or "adaptability." In this sense, then, general collective intelligence means "group versatility" or "group adaptability," and a supermind is a versatile or adaptable group.

What Kind of Intelligence Do Computers Have?

The distinction between specialized intelligence and general intelligence helps clarify the difference between the abilities of today's computers and human abilities. Some artificially intelligent computers are far smarter than people in terms of certain kinds of specialized intelligence. But one of the most important things most people don't realize about AI today is that it is all very specialized.[1]

Google's search engine is great at retrieving news articles about baseball games, for example, but it can't write an article about your son's Little League game. IBM's Watson beats humans at *Jeopardy!*, but the program that played *Jeopardy!* can't play tic-tac-toe, much less chess.[2] Teslas can (sort of) drive themselves, but they can't pick up a box from a warehouse shelf.

Of course, there are computer systems that can do these other things. But the point is that they are all different, specialized programs, not a single general AI that can figure out what to do in each specific situation. Humans, with their general intelligence, must write programs that contain rules for solving different specific problems, and humans must decide which programs to run in a given situation.

In fact, none of today's computers are anywhere close to having the level of general intelligence of any normal human 5-year-old. No single computer today can converse sensibly about the vast number of topics an ordinary 5-year-old can, not to mention the fact that the child can also walk, pick up weirdly shaped objects, and recognize when people are happy, sad, or angry.

How soon, if ever, will this change? Progress in the field of artificial intelligence has been notoriously difficult to predict ever since its early days in the 1950s. When researchers Stuart Armstrong and Kaj Sotala analyzed 95 predictions made between 1950 and 2012 about when general AI would be achieved, they found a strong tendency for both experts and nonexperts to predict that it would be achieved between 15 and 25 years in the future—regardless of when the predictions were made.[3] In other words, general AI has seemed about 20 years away for the last 60 years.

More recent surveys and interviews tend to be consistent with this long-term pattern: People still predict that general AI will be here in about 15 to 25 years.[4] So while we certainly don't know for sure, there is good reason to be skeptical of confident predictions that general AI will appear in the next couple of decades. My own view is that, barring some major societal disasters, it is very likely that general AI will appear *someday*, but probably not until quite a few decades in the future.

All uses of computers will need to involve humans in some way until then. In many cases today, people are doing parts of a task that machines can't do. But even when a computer can do a complete task by itself, people are always involved in developing the software and usually modifying it over time. They also decide when to use different programs in different situations and what to do when things go wrong.

How Can People and Computers Work Together?

One of the most intriguing possibilities for how people and computers can work together comes from an analogy with how the human brain is structured. There are many different parts of the brain that specialize in different kinds of processing, and these parts somehow work together to produce the overall behavior we call intelligence. For instance, one part of the brain is heavily involved in producing language, another in understanding language, and still another in processing visual information. Marvin Minsky, one of the fathers of AI, called this architecture a "society of mind."[5]

Minsky was primarily interested in how human brains worked and how artificial intelligence programs might be developed, but his analogy also suggests a surprisingly important idea for how superminds consisting of both people and computers might work: Long before we have general AI, we can create more and more collectively intelligent systems by building societies of mind that include both humans and machines, each doing part of the overall task.

In other words, instead of having computers try to solve a whole problem by themselves, we can create cyber-human systems where multiple people and machines work together on the same problem. In some cases, the people may not even know—or care—whether they are interacting with another human or a machine. People can supply the general intelligence and other skills that machines don't have. The machines can supply the knowledge and other capabilities that people don't have. And, together, these systems can act more intelligently than any person, group, or computer has done before.

How is this different from current thinking about AI? Many people today assume that computers will eventually do most things by themselves and that we should put "humans in the loop" in situations where people are still needed.[6] But it's probably more useful to realize that most things now are done by groups of people, and we should put computers into these groups in situations where that is helpful. In other words, we should move away from thinking about *putting humans in the loop* to *putting computers in the group*.

What Roles Will Computers Play Relative to Humans?

If you want to use computers as part of human groups in your business or other organization, what roles should computers play in those groups? Thinking about the roles that people and machines play today, there are four obvious possibilities. People have the most control when machines act only as tools; and machines have successively more control as their roles expand to assistants, peers, and, finally, managers.

Tools   A physical tool, like a hammer or a lawn mower, provides some capability that a human doesn't have alone—but the human user is directly in control at all times, guiding its actions and monitoring its progress. Information tools are similar. When you use a spreadsheet, the program is doing what you tell it to do, which often increases your specialized intelligence for a task like financial analysis.

But many of the most important uses of automated tools in the future won't be to increase individual users' specialized intelligence, but to increase a group's collective intelligence by helping people communicate more effectively with one another. Even today, computers are largely used as tools to enhance human communication. With email, mobile applications, the web in general, and sites such as Facebook, Google, Wikipedia, Netflix, YouTube, and Twitter, we've created the most massively connected groups the world has ever known. In all these cases, computers are not doing much "intelligent" processing; they are primarily transferring information created by humans to other humans.

While we often overestimate the potential of AI, I think we often underestimate the potential power of this kind of hyper-connectivity among the 7 billion or so amazingly powerful information processors called human brains that are already on our planet.

Assistants A human assistant can work without direct attention and often takes initiative in trying to achieve the general goals someone else has specified. Automated assistants are similar, but the boundary between tools and assistants is not always a sharp one. Text-message platforms, for instance, are mostly tools, but they sometimes take the initiative and autocorrect your spelling (occasionally with hilarious results).

Another example of an automated assistant is the software used by the online clothing retailer Stitch Fix to help its human stylists recommend items to customers.[7] Stitch Fix customers fill

out detailed questionnaires about their style, size, and price preferences, which are digested by machine-learning algorithms that
select promising items of clothing.

The algorithmic assistant in this partnership is able to take
into account far more information than human stylists can. For
instance, jeans are often notoriously hard to fit, but the algorithms are able to select for each customer a variety of jeans that
other customers with similar measurements decided to keep.

And it is the stylists who make the final selection of five items
to send to the customer in each shipment. The human stylists
are able to take into account information the Stitch Fix assistant
hasn't yet learned to deal with—such as whether the customer
wants an outfit for a baby shower or a business meeting. And,
of course, they can relate to customers in a more personal way
than the assistant does. Together, the combination of people and
computers provides better service than either could alone.

Peers   Some of the most intriguing uses of computers involve
roles in which they operate as human peers more than assistants
or tools, even in cases where there isn't much actual artificial
intelligence being used. For example, if you are a stock trader,
you may already be transacting with an automated program
trading system without knowing it.

And if your job is dealing with claims for Lemonade Insurance Agency, you already have an automated peer named AI
Jim.[8] AI Jim is a chatbot, and Lemonade's customers file claims
by exchanging text messages with it. If the claim meets certain
parameters, AI Jim pays it automatically and almost instantly.
If not, AI Jim refers the claim to one of its human peers, who
completes the job.

Managers Human managers delegate tasks, give directions, evaluate work, and coordinate others' efforts. Machines can do all these things, too, and when they do, they are performing as automated managers. Even though some people find the idea of a machine as a manager threatening, we already live with mechanical managers every day: A traffic light directs drivers; an automated call router delivers work to call center employees. Most people don't find either situation threatening or problematic.

It's likely that there will be many more examples of machines playing the role of managers in the future. For instance, the CrowdForge system crowdsources complex tasks such as writing documents. In one experiment, the system used online workers (recruited via the Amazon Mechanical Turk online marketplace) to write encyclopedia articles.[9] For each article, the system first asked an online worker to come up with an outline for the article. Then it asked other workers to find relevant facts for each section in the outline. Next it asked still other workers to write coherent paragraphs using those facts. Finally, it assembled the paragraphs into a complete article. Interestingly, independent readers judged the articles written in this manner to be better than articles written by a single person.

How Can Computers Help Superminds Be Smarter?

If you want to design a supermind (like a company or a team) that can act intelligently, it needs to have some or all of the five cognitive processes that intelligent entities have—whether they are individuals or groups. Your supermind will need to create possibilities for action, decide which actions to take, sense the external world, remember the past, and learn from experience.

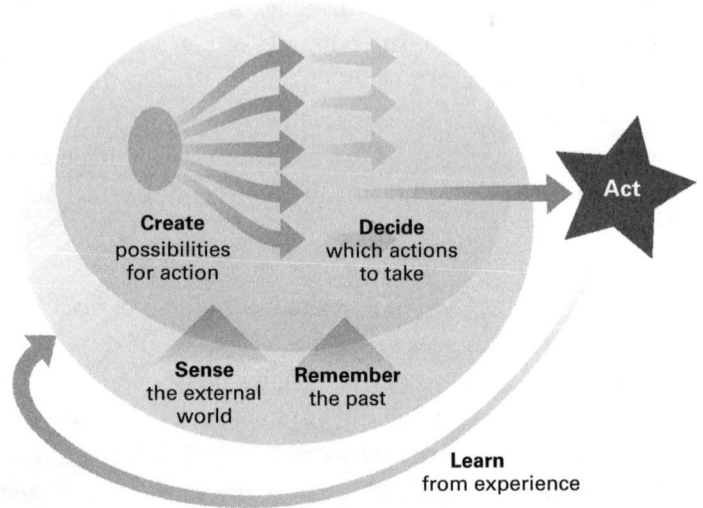

The Basic Cognitive Processes Needed by Any Intelligent Entity
Entities that act intelligently (such as people, computers, and groups) need to do these things.

Computers can help do all these things in new ways that often—but, of course, not always—make the superminds smarter. To see how, let's consider how a large corporation like Procter & Gamble could develop a new strategic plan. The possibilities we'll discuss are just that: possibilities. I have no reason to believe that P&G is doing these things at present. But I think that P&G and many other companies are likely to do things like this in the future.

Today, corporate strategic planning in large companies usually involves a relatively small group of people, mostly senior executives, their staff, and perhaps some outside consultants. But what if we could use technology to involve far more people and let machines do some of the thinking?

Create As we saw above, one of the most important roles for computers is as a communication tool that allows much larger groups of people to think together productively. A promising approach for doing that within the strategic planning process is to use a family of related online contests, called a contest web.[10] There could be separate online contests for strategies at different levels of the organization. For example, if P&G used this approach, the company might have separate contests for each brand, such as Pantene shampoo, Head & Shoulders shampoo, and Tide laundry detergent. It could also have separate contests for how to combine the strategies of the brands in each business unit, such as hair care and fabric care. And the company could have another contest aimed at combining the business unit strategies into an overall corporate strategy.

Each contest could be open to many company employees, perhaps all of them. Anyone in the contest could propose a strategic option, and others could comment on or help develop the idea. Eventually there would be one winning strategy chosen in each challenge, but during the planning process, it would be important to consider a number of different options.

Opening this process to lots of people could allow surprising new options to arise. For instance, a group of young, tech-savvy employees, who would never have been included in a traditional corporate strategic-planning process, might propose a new cosmetics concept involving skin and eye makeup specially formulated for individual customers who upload selfies to the website.

Decide One benefit of involving more people in generating strategic possibilities is that you get far more possibilities. But deciding which possibilities are most promising requires evaluating them all, and new technologies also make it easier to involve far more people and more kinds of expertise in evaluating. For instance, P&G might want its manufacturing engineers to evaluate whether it is technically feasible to make a product, its operations managers to estimate manufacturing cost, and perhaps outside market researchers to predict the demand for the product at different price points.

In some cases, it may be worth combining many people's opinions about some of these questions. For instance, P&G might use online prediction markets to estimate the demand for products. Such markets have already been used to successfully predict movie box-office receipts, winners of U.S. presidential elections, and many other things. Somewhat like futures markets, prediction markets let people buy and sell "shares" of predictions about future events. For instance, if you believe that global sales for Pantene shampoo will be between $1.8 billion and $1.9 billion per year, you could buy a share of this prediction. If the prediction is right, then you will get, say, $1 for each share you own of that prediction. But if your predictions are wrong, you will get nothing.[11] That means the resulting price in the prediction market is essentially an estimate of the probability that sales will be in this range.[12]

Sense A key necessity for developing good strategic plans is the ability to effectively sense what is going on in the external world: What do customers want now? What are our competitors doing? What new technologies might change our industry? By far the most visible means for improving sensing today are big data and data analytics.

For example, P&G might analyze the positive and negative comments about its products in online social networks to gauge how customer sentiment about the products is changing. It might conduct online experiments at different prices for the products. And it might be able to obtain early warnings about sales changes by installing video and touch-sensitive floors in retail stores to analyze how much time customers spend looking at P&G's products versus competitors' products.

P&G might even be able to do something Amazon has already done: use vast amounts of data to develop detailed models of many parts of its business, such as customers' responses to prices, ads, and recommendations, and sort out how supply-chain costs vary with inventory policies, delivery methods, and warehouse locations.[13] With tools like these, computers can take over much of the quantitative work of strategic planning by running the numbers, and people can use their general intelligence to do more qualitative analysis.

Remember Another way technology can help superminds create better strategic plans is by helping them remember good ideas that others have had in similar situations. For example, software assistants embedded in an application for generating strategy proposals could automatically suggest generic strategies, such as the following:

- Integrating forward by taking on some of the tasks done by your customers, or integrating backward by taking on some of the tasks done by your suppliers;

- Outsourcing more of the things you do internally to freelancers or specialized providers;

- Moving into related market segments, nearby geographical regions, or other markets frequented by your customers.

When you pick one of these options, the system could then automatically provide a template including the kinds of details necessary for that type of strategy.

By remembering good strategies from other settings, software assistants could help generate new strategies for your setting. For instance, if the strategy of using selfies to customize cosmetics were successful, a software assistant could suggest similar strategies that let customers use smartphones to customize P&G's other products: shampoos, toothpastes, laundry detergents, potato chips, and others. Of course, many of these combinations would be silly or impractical and could be very quickly eliminated, but some might be surprisingly useful. And even silly options sometimes give rise to good ideas.

For instance, in the early 2000s, P&G developed a process for printing entertaining pictures and words on Pringles potato chips.[14] An approach like this might have led to another promising idea: using this technology to let customers buy Pringles that are preprinted with images that customers specify themselves.

Learn If a system is used over time, it can help a supermind learn from its own experience to become more and more effective. For example, it might help recognize strategic ideas that most people wouldn't recognize in their early stages. In the 1970s, when Steve Jobs and Bill Gates were first playing around with what we now call personal computers, most people had no idea that these strange, awkward devices would turn out to be among the most innovative and influential products of the next several decades.

It's certainly not easy to rapidly filter ideas without missing these diamonds in the rough. But perhaps it's possible to identify the unusual people who do have this skill by systematically tracking over time how accurately, and how early, people predict technological advances and other kinds of breakthroughs. Then we could ask these people to take a second look at some of the "crazy" ideas that we might otherwise reject.

Another intriguing possibility is to use "learning loops" that begin with human experts evaluating strategies manually and then gradually automate more and more of the work as the machines get better at predicting what human experts would do.

In a company like P&G that generally tries to compete on quality rather than price, experts evaluating product strategies usually reject those that emphasize competing on low price. But instead of programmers writing programs that explicitly filter out low-price strategies, a machine-learning program might recognize that experts often reject these types of strategies and start suggesting this action. If the experts agree with the suggestion enough times, then the program might stop asking and just do the filtering automatically.

A Cyber-Human Strategy Machine

You might call the kind of strategic planning process I've described above a cyber-human strategy machine.[15] Given how complex such a system could be and how generic much of the work would be, it seems unlikely that companies would develop proprietary systems for this purpose. Instead, today's consulting firms, or their future competitors, might provide much of this functionality as a service. Such a strategy-machine company, for instance, could have a stable of people at many levels of expertise on call who could rapidly generate and evaluate various strategic possibilities, along with software to automate some parts of the process and help manage the rest.

In the long run, such a strategy machine might use a super-mind of people and computers to generate and evaluate millions of possible strategies for a single company. Computers would do more and more of the work over time, but people would still be involved in parts of the process. The result would be a handful of the most promising strategic options that the human managers of the company would choose among.

The examples we've just discussed are focused on strategic decision making, but what we've really seen is an architecture for general purpose, problem-solving superminds: Computers use their specialized intelligence to solve parts of the problem, people use their general intelligence to do the rest, and computers help engage and coordinate far larger groups of people than has ever been possible.

As new technologies make this easier, we are likely to see many more examples of human–computer superminds being used to solve all kinds of business and societal problems—not just corporate strategic plans, but also designs for new houses, smartphones, factories, cities, educational systems, antiterrorism approaches, and medical treatment plans. The possibilities are virtually unlimited.

7

Face the Future of Work

Lynda Gratton

We are living through a grand transition in the way people work. Constant and extraordinary innovation in machine learning and robotics is reshaping our professional lives. Some tasks will be replaced. Others will be augmented. No one—not even the highly skilled—will be untouched.

Meanwhile, as people live longer and their working lives expand, they are moving from the traditions of the three-stage life—full-time education leading to full-time work leading to full-time retirement—to something a great deal more fluid, flexible, and multistaged. And as more women work and more partnerships are built on "career plus career" rather than "career plus carer," technological and demographic forces are further altering the relationship employees and their families have with the workplace.

My interest here is what all this means for leaders.

In many ways, leaders' day-to-day lives are more protected from these major shifts than those of many employees. The complexity of a leader's work makes positive augmentation rather than replacement the most likely outcome of technological innovation. With their capacity for wealth creation, leaders have more opportunities to "go plural" and work with multiple organizations at the same time and to develop encore careers with relative ease. Perhaps as a result of their own protection within the workplace, some leaders have failed to realize that the daily lives of those who work in their organizations will inevitably be transformed over the coming decades.

But leaders need to be deeply aware—right now, not down the line—of the transition taking place. And they need to have a clear sense of what they can do to prepare their employees for the future of work.

How Leaders Can Pave the Way

Specifically, leaders must take the following three steps:

Create a narrative about the future of jobs   Most people have an idea that technology will alter their work. Some, such as drivers, cashiers, and salespeople, know that this will occur in the near term. We know from employee surveys that these changes are a concern: Many workers fret that they don't have the right skills to do new jobs, and few even know what these new jobs could be. Really understanding the impact of technology on jobs requires a fine-grained analysis of the country, the sector, the job, the tasks, and the skills; there are no easy answers.

So it becomes the role of an organization's leader to create a narrative about future job pathways and likely job creation. This does not have to be, and in many cases cannot be, precise. The leader's narrative should acknowledge that work is changing and offer a job-by-job analysis that provides some idea of what this trajectory could look like. Employees can then engage their own sense of agency and motivation to think about how they can take action.

Support learning One of the biggest changes arising from the intersection of technological innovations and increasing longevity is that one-off early education will not be sufficient to propel people through their whole working lives. People will need to engage in work that has development opportunities built into it, be prepared to spend some of their leisure time upskilling, and most likely take significant chunks of time out of work to learn new skills.

Many companies shy away from this responsibility, believing that in a volatile labor market with short job tenure, it is not in their interest to help develop employees.

This is the wrong approach. Increasingly, people will choose companies with the capacity to create learning opportunities and will stay and flourish because such opportunities are available. So the second way leaders should prepare for the future of work is to champion a learning agenda by prioritizing their involvement in learning initiatives and by modeling adult learning through their own development activities.

Model flexible working As jobs and skill requirements change, so will the ebb and flow of daily life. To understand this, imagine for a moment that you are in a job that you know technology will transform and you are keen to learn and develop. Imagine also that you believe you will live into your 90s, maybe over 100, and you have calculated that to do so, you will need to work into your mid-70s or longer. This is a calculation that many people will be making. Multistaged lives will take many forms: Some people will take time out to explore new directions in their 40s, some will work part-time in their 50s while bringing up children, and some will jump back into serious work in their 60s, when they still have lots of energy and there's more time to focus on work.

In these scenarios the 9-to-5 workday (or 8-to-8 in some sectors), the five-day workweek, and the limited holiday entitlement seem hopelessly misaligned. Some companies have realized this and have created greater opportunities for flexible work, job sharing, paternity leave, sabbaticals, and midlevel hires.

But here's the rub: People are disinclined to take advantage of many of these opportunities for fear that doing so will adversely affect their careers, signaling that they are not aligned with the values of a high-performing culture. Leaders must address this fear by modeling flexible working themselves. That is one of the most crucial ways a leader can prepare employees for the future of work. Otherwise, people won't believe that it's safe to work flexibly. If the boss does it, they'll feel freer to do it, too.

Don't Let Fear Take Root

Many of us are worried about the future of work. When we feel stressed, we are less likely to be cooperative, less likely to innovate, and more likely to be aggressive.

Leaders can help alleviate all that by narrating how the future could be, by supporting learning, and, perhaps most important, by modeling flexibility. Decades from now, such leaders will be remembered as those who were best prepared to navigate this time of intense transitions—and who did not let fear take root.

8

How Emotion-Sensing Technology Can Reshape the Workplace

Eoin Whelan, Daniel McDuff, Rob Gleasure, and Jan vom Brocke

As companies search for new ways to improve performance, some executives have begun paying attention to developments in emotion-sensing technologies (ESTs) and software fueled by artificial emotional intelligence. Although we are still in the early days, research shows that these technologies, which read such things as eye movements, facial expressions, and skin conductance, can help employees make better decisions, improve concentration, and alleviate stress. While important privacy issues need to be addressed, the opportunities are significant.

Consider the technology developed by technology multinational Philips and Dutch bank ABN AMRO to reduce trading risk in financial markets. Research has shown that traders in heightened emotional states will overpay for assets and downplay risk, a condition known as "auction fever" or "bidding frenzy." To address this problem, the companies jointly developed a tool called the Rationalizer that has two components: a bracelet attached to the trader's wrist that measures emotions via electrodermal activity (similar to the way a lie detector works) and a display showing the strength of the person's emotions using light patterns and colors. Researchers have found that when users become aware of their heightened emotional states, they are more likely to rethink their decisions. In addition to helping individuals improve performance, the aggregated data from such settings can help managers understand how internal and external environmental factors influence the risks taken by groups.

Individuals are also more prone to make mistakes when they are not paying enough attention. Although multitasking has become standard in many jobs, there are some activities, such as air-traffic control and fast-paced buying and selling, where maintaining one's undivided attention is critical. In a high-profile foul-up in 2005, a trader working for Mizuho Securities intended to sell a single share of a stock it owned for about 610,000 yen (which was approximately $5,000). By mistake, he placed an order to sell 610,000 shares for one yen. The company was unable to cancel the sell order, leading to an estimated loss of $224 million.

Although such egregious blunders are rare, the story speaks to how important it is to hold the attention of employees involved in high-stakes activities. ESTs can help people improve their focus, often with relatively minimal technological investment. For example, recent research has found that slow or uneven cursor movements can be an indication of distraction or negative emotions. Detection doesn't require installing expensive hardware, but rather just some additional code or software to computers or smartphones.

Professional athletes have been early adopters of tools that can help people sharpen their focus to gain a competitive edge. Major League Baseball All-Star Carlos Quentin, National Basketball Association All-Star Kyle Korver, and Olympic gold medal swimmer Eric Shanteau are among those who have used special headsets produced by SenseLabs to monitor cognitive

In some organizations, HR departments try to monitor stress levels using surveys. But surveys don't necessarily capture how employees actually feel.

performance and develop customized training aimed at shoring up their personal weaknesses. Microsoft has also conducted research on the use of wearable sensors in an effort to understand, among other things, what work activities are associated with changes in emotion and when people working on certain types of tasks should take breaks.

In settings where employee engagement is critical, the ability of managers to recognize boredom is vital. Assuming that data can be accessed without compromising privacy or anonymity, managers will soon be able to watch for signs of boredom in an underperforming team and take steps to counter it. Indeed, researchers at Telefónica I+D have developed an algorithm that analyzes smartphone activity for such signs. On the basis of a combination of data points—including how often users check their email, whether they log in to Instagram, whether they are adjusting their device settings, and how much battery power they consume—the algorithm correctly identifies user boredom more than 80% of the time. It can tell when employees use their phones to pass the time as opposed to pursuing specific goals.

In light of such discoveries, managers can seek to redesign processes that induce boredom or alternate them with other activities that employees find more engaging. ESTs, moreover, might help managers figure out which work schedules work best for particular teams: Employees in one group may be most productive in the early morning, while another group may do better later in the day. Meeting schedules could be organized to take advantage of this information.

Reducing Stress and Burnout

Although some types of stress can help people focus, research shows that too much stress is detrimental to productivity, creativity, and job satisfaction, not to mention psychological and physical health. What's more, stress can reach harmful levels long before people are aware of it. In some organizations, human resources departments try to monitor stress levels using surveys. But surveys don't necessarily capture how employees actually feel, in part because people don't always know when their stress levels are elevated. Having a tool that provides a quantifiable, objective measure of stress would be extremely helpful.

As with tools to improve decision making and focus, numerous options are available, including smart watches and fitness trackers that detect stress by measuring changes in heart rate and sweat (through what's known as electrodermal activity). These measures can identify small changes that users themselves don't notice. And as with algorithms that monitor smartphone usage for boredom or cursor activity for distraction, stress-related information can also be drawn from the hardware that people are accustomed to using every day. For example, a study by MIT's Affective Computing Lab found that computer users who were under stress pushed harder on keyboard keys and held the mouse more tightly. Other research has found that it's possible to detect stress-related surges in heart rates by monitoring the changes in the light reflected off users' faces with an ordinary webcam.

We have found that there can be important benefits to monitoring stress at both the individual level and across the organization. At the individual level, managers can learn when people are under sustained pressure (and therefore more susceptible to recklessness, burnout, or conflict with others) and take steps to help ameliorate such situations. At an organizational level, measuring physiology (for example, heart rate or electrodermal activity) can help managers identify stress "hot spots" among teams and functions. Using wristbands or webcams, for example, managers can pick up on problems relating to excessive workload or interpersonal conflict and respond to them, often before employees are aware they exist. Employees may be spinning their wheels on frustrating, unproductive activities (for example, arguing over who has responsibility for specific tasks). Having access to this data might allow managers to create a "heat map" indicating where the problem is concentrated.

Addressing the Barriers

As companies become interested in ESTs, they will need to address barriers related to cost, complexity, and issues of privacy.

The cost- and complexity-related barriers seem to be relatively straightforward—both have been declining, and numerous low-cost/low-complexity options are already available. Allaying the privacy concerns, however, will be trickier. Many employees are highly skeptical of monitoring technology and uneasy about how ESTs might be used. A fundamental issue is who will get to see the data and whether the data will be broken down individually or aggregated across groups. Such concerns are understandable given that much of the value will come from measuring and managing aspects of behavior that people are unable (or perhaps unwilling) to self-report. Even if all parties agree to common rules for consent, anonymity, and personal well-being, there are lingering issues. For example, what happens if ESTs uncover medical issues that individuals aren't aware of or wish to keep private?

One can speculate that privacy concerns will become less problematic when the people being measured are the beneficiaries and when disclosure is voluntary. But even then, there are

Implementation Barriers for Emotion-Sensing Technologies

Multiple measures are available to appraise stress, attention, and decision making. Each presents different cost-, complexity-, and privacy-related barriers.

Organization opportunity	Relevant measures	Cost-related barriers	Complexity-related barriers	Privacy-related barriers
Decision making	Blood sugar	Low-medium	Low	Medium
	Electrodermal	Medium	Low	Medium
	EEG	High	Medium	High
Attention	Mouse/browser tracking	Low	Medium	Low-medium
	Facial coding	Medium	Low	Medium
	Eye tracking	Medium-high	Low	Low-medium
Stress	Hormones	Low	Low	Low-medium
	Heart rate	Low-medium	Medium	Medium
	Electrodermal	Medium	Low	Medium-high

dicey issues, such as whether an employee interprets feedback in an unexpected way or overadjusts to correct behaviors. With that in mind, managers can attempt both to maintain oversight and to reduce employee concerns by doing the following:

1. **Be sensitive to employee concerns.** Prepare your organization for using ESTs through education and transparency. Explain how the tools can benefit employees by reducing stress and risks of burnout. One potentially useful strategy, known as BYOD, involves inviting employees to bring their own devices to work. Under this scenario, individuals maintain a sense of ownership over the deployment of ESTs and the data they are gathering.

2. **Develop data governance agreements.** Employees should have sole control over their personal emotional data and be able to stipulate what types of usage are permitted (for example, data can be used only on an aggregate level, and no one can drill down into individual data signatures).

3. **Similarly, assure employees in written agreements that emotional data will be used only for specific business goals.** For technologies that rely on broad-stroke measures, such as webcam-based emotion detection, data gathering and analysis should be directed toward highly specific and well-defined outcomes.

As long as organizations operate responsibly, we believe employees will gradually become comfortable with the gathering and analysis of physiological, behavioral, and emotional data. Although this won't happen overnight, several trends suggest that trust can be built over time. Millions of individuals already use smart watches and fitness devices like Apple Watches and Fitbits, and many people share their workout and nutrition data openly on social media. Social media itself has conditioned us to accept and even embrace new levels of personal transparency. The challenge will be to introduce new devices and measures into workplaces in a way that empowers performance, mitigates privacy concerns, and generally reassures employees that the benefits are mutual.

9

When Communication Should Be Formal

Antti Tenhiälä and Fabrizio Salvador

When we communicate in organizations, we tend to keep things casual so that we can be fast and flexible and get things done. We email, Skype, Slack, and Yammer. Formal, protocol-guided communication—such as face-to-face meetings or teleconferences, where leaders use standard agendas to review concerns and coordinate responses—is increasingly seen as an old-fashioned bureaucratic time sink.

Informality helps an organization's daily operations run more smoothly, to be sure. And unnecessary meetings that serve no real business purpose can plague a workplace. But no one would argue against the value of formal, reliable communication in, say, aviation or the military. In those mission-critical contexts, protocols for a communication's timing, content, and participants ensure clarity, transparency, and accountability.

Prompted by those models, we decided to study companies that manufacture high-tech machinery—businesses that need precise, cross-functional communication to get the job done. Our data shows that overreliance on informal communication can harm performance because it is often imprecise and erratic, and that formal communication offers specific, crucial advantages that no company should overlook.

Over the course of two years, we studied 73 manufacturing sites encompassing 163 production processes for customized industrial machinery and instruments. We analyzed both ongoing informal communication (such as emails and phone calls) and periodic (typically weekly) cross-functional meetings with standard agendas and prespecified participants. On-time delivery—a critical performance dimension in this industry—was our primary measure of communication effectiveness. We found that processes using periodic protocol-guided meetings had a consistent performance advantage over those relying solely on informal communication. Indeed, they improved the rate of on-time delivery by an average of 5 to 8 percentage points, representing substantial value for the companies we studied.

About the Research

This chapter is based on a two-year research project conducted at high-tech machinery manufacturers in 18 countries across Europe, Asia, and North America. We analyzed formal and informal communications among managers from sales, product design, engineering, production, and procurement. Operational communications in this industry often center on customer-initiated changes to order specifications and delivery dates, engineering modifications to product designs, resource availability issues in production, and the logistics of shipments from suppliers.

Our primary performance measure was on-time delivery rate. Secondary measures included the ability to promise delivery by customers' initially requested dates and speed of order fulfillment. In comparing the influence of formal versus informal communication on these measures, we statistically controlled for other major influences on delivery performance, such as raw materials delays, internal quality issues, machine breakdowns, and delays caused by customers.

Overall, use of formal meeting protocols (rather than solely informal communications) to address product specification changes improved the rate of on-time delivery by, on average, 5 percentage points. When the communications were about customer-requested delivery-date adjustments, use of formal protocols improved on-time delivery by an average of 8 percentage points. The advantages of formal communication were statistically significant—and similar in magnitude between the primary and secondary performance measures.

Recognizing the Value of Formality

Only 45% of the organizations in our study relied on formal meeting protocols. The most widely used channel of operational communication was email (used by 71% of the organizations). When we interviewed managers, they often said that they chose email primarily because it offered speed and flexibility—and that they opted against recurring meetings mainly because staff resisted them. Informal channels are indeed speedier and more flexible than formal communication, and they can be useful when the matter at hand is truly novel or complex enough to merit a rapid back-and-forth discussion. But when you're in the weeds of daily operations, tasks can seem more novel and complex than they actually are. Furthermore, if you opt for an informal exchange, you risk connecting with the wrong people (perhaps because the right people are not readily available), delivering or receiving inaccurate or incomplete messages, and getting distracted from the current interaction by competing priorities. As a result, reliance on informal communication often leads to delays, rework, contract penalties, costly expediting efforts, and disappointed customers.

For example, we observed that critical messages were some-times held up or even forgotten because stakeholders did not immediately respond to one another's ad hoc requests. In other cases, individuals sought guidance on decisions but—after their informal messages ricocheted around their organizations—moved forward on their own, having never received a clear yes or no or a proper assessment of downstream implications.

Establishing a protocol takes effort and necessitates overcom-ing the common assumption that formality means drudgery and inertia. But in settings where communication errors can be costly, formal protocols can be a rock of reliability for the fol-lowing reasons:

- They allow people to connect with the right stakeholders at the right time.
- They standardize messages to ensure that they are complete, and provide standardized procedures for follow-up.
- They promote accountability for tasks, because responsibili-ties are explicitly assigned to specific people.
- They embed lessons from previous interactions and meetings to ensure continuous improvement.

The health care sector is learning this lesson. According to The Joint Commission, the largest medical services accredita-tion agency in the United States, up to 80% of serious medical errors stem from miscommunication among caregivers.[1] Patient handoffs between intensive care units and operating rooms, for example, are essentially cross-functional meetings that demand precise exchanges of information. It's not surprising that hospi-tals are making substantial efforts to improve communication in such instances, and standardized protocols have proved to be an effective means of doing that.[2] Leading health care providers, such as the University of Pennsylvania Health System, have developed stepwise protocols to standardize what these exchanges cover, which parties should be present, and who is responsible for transmitted information.[3]

Designing Communication Protocols

It is not necessary to set up formal protocols for communication regarding infrequent events. Informal channels are just fine for that purpose. For example, if a member of a procurement team must contact a marketing expert—an interaction we seldom observed in our research—it makes sense to use an enterprise social network platform to identify and connect with an appropriate person. To establish a protocol for that rare type of communication would be a waste of time.

However, formal communication is especially effective for common events. For instance, if a procurement team learns that a shipment from a supplier will be delayed—an event we observed frequently in our research—a formal communication protocol ensures that accurate information about the situation reaches the right production planners and sales reps so that the timing of orders can be adjusted and customers can be notified.

When events are sufficiently frequent, the organization learns who should be contacted under what circumstances— lessons that a protocol can codify. The organization also identifies which errors and which types of miscommunication are most common, and a well-designed protocol can address those sticking points.

But why make a protocol for something that is already done repetitively? Again, a useful lesson comes from health care—specifically, from University of Iowa Hospitals & Clinics. A new cross-departmental communication protocol that standardized patient handoffs between emergency departments and operating rooms improved the quality of care while radically reducing lead times. "There are some things that you never think to plan for, especially things that you do every day," concluded the team of physicians behind the new protocol. "They may seem too trivial or common to really organize, and afterward you often think that it could have gone better."[4]

Afterthoughts like that are invaluable input for the development of communication protocols. And such protocols ideally should be created jointly by people who collaborate regularly. When individuals develop communication habits on their own, their assumptions about best practices often do not align with those of their colleagues, and that can lead to frustrating situations fraught with competing expectations and unsynchronized efforts.

Overcoming Resistance

Getting people to develop and commit to a formal communication protocol is a considerable challenge. After all, most people prefer to craft messages in their own words and send them to whomever they regard as relevant at whatever time they choose.

Addressing the negative perceptions of formality is therefore crucial for motivating people to implement change. In some organizations, it may be effective to hold up agile management methodology as a positive example. Agile has come a long way from its software development origins, with contemporary implementations ranging from banking[5] to boardrooms.[6] Although the approach emphasizes self-organization, at its core is a strict communication protocol for team meetings called "scrums."[7]

The scrum format is always the same (with most colocated teams standing up instead of sitting down). The participants and timing constraints are prespecified, and the agenda is limited to three questions: What has everyone done since the previous meeting? What will they do next? What is everyone's most pressing challenge? Scrum may not be the perfect communication protocol for every organization, but it has the potential to appeal to people who bristle at anything that seems old-fashioned and bureaucratic.

A common argument against formality is that even frequent events have their own particularities that demand informal, unique communications. Take, for example, customers' changes to orders, a common challenge in capital goods manufacturing. Sales staff routinely claim that their valued customers deserve fast, individually tailored responses to their requests for changes, and it may seem impossible to respond quickly when using formal communication channels.

Informal communication may indeed get fast attention, but it is prone to losing attention just as quickly. If all the relevant information is not captured in the moment, and if conversations are forgotten, misunderstandings and mistakes may arise, and customer dissatisfaction and delivery delays increase. For example, the immediate gratification of swiftly confirming a customer order amendment may be undercut by the later discovery that the new delivery date cannot be met—because not all relevant parties were consulted or because follow-up was inadequate to ensure thorough, coordinated implementation of changes.

Collecting and presenting data on past miscommunications— and the resulting delivery failures—may help mitigate resistance to the adoption of formal communication protocols. In our study, formal communication was more common in organizations that approached process improvement systematically (sometimes with Six Sigma). Of the many benefits of data-driven problem solving, one emphasized by managers we interviewed is the simple fact that hard numbers can be irresistibly persuasive. For example, one process at a robotics manufacturer we studied had unsatisfactory delivery performance. Although everyone was aware of the problem, improvement efforts did not initially focus on shortcomings in communication. A shift in focus did not occur until analysis of hard data showed that internal confusion about product-specification changes caused delivery failures more often than any external factor. Of course, such analysis alone is insufficient to prove the effectiveness of formal communication, but it does suggest a need for change.

Once a formal protocol has been piloted, comparative analyses can provide impetus for a wider rollout. One such analysis, at a manufacturer of industrial refrigeration systems, revealed that customer order amendments were twice as likely to result in delivery failures—and that their average cost doubled—when changes were communicated via email rather than in weekly meetings between production planners and sales personnel. Faced with stark numbers, even the most reluctant sales reps admitted that the convenience of email just didn't matter.

Formality Forward

Our aim is not to urge organizations to return to the days when every process was documented in triplicate or to make people jump through hoops just for the sake of uniformity. It is merely to show, with evidence from our research, that informal communication has its limits and should not be blindly accepted as a best practice. Lessons from aviation, the military, health care—and now high-tech manufacturing—reveal that formal communication (like organizational hierarchy[8]) not only has a place in everyday operations but also offers competitive advantages that no forward-looking company can afford to ignore.

III
___

How We Manage

10

Improving Communication in Virtual Teams

N. Sharon Hill and Kathryn M. Bartol

As collaborative technologies proliferate, it is tempting to assume that more sophisticated tools will engender more effective virtual communication. However, our study of globally dispersed teams in a major multinational organization revealed that performance depends on how people use these technologies, not on the technologies themselves.

We asked team members to rate one another on virtual communication behaviors culled from a growing body of research on virtual teams. Peer assessments focused on five best practices: matching the technology to the task, making intentions clear, staying in sync, being responsive and supportive, and being open and inclusive. (Participants had worked together for some time and had been tasked with improving key business processes.) Individual scores were averaged to determine team scores.

When controlling for past experience on virtual teams and level of technology support available, we found that teams with higher scores on the five behaviors also received higher ratings from their leaders on producing quality deliverables, completing tasks on time, working productively together, and meeting or exceeding goals. Results indicated a linear relationship across the board: For every 10% that a team outscored other teams on virtual communication effectiveness, they also outscored those teams by 13% on overall performance. Although the research focused on dispersed teams, we believe the same strategies can help colocated teams, which increasingly depend on virtual collaboration tools.

Let's look at each of the five behaviors in detail. They may seem basic at first glance, but we've observed that they are often overlooked. When teams are informed of these simple strategies and take steps to implement them, they outperform teams that don't.

Match the technology to the task   Teams have many communication technologies at their disposal, ranging from email and chat platforms to web conferencing and videoconferencing. People often default to using the tool that is most convenient or familiar to them, but some technologies are better suited to certain tasks than others, and choosing the wrong one can lead to trouble.

Communication tools differ along a number of dimensions, including information richness (or the capacity to transfer nonverbal and other cues that help people interpret meaning) and the level of real-time interaction that is possible. A team's communication tasks likewise vary in complexity, depending on the need to reconcile different viewpoints, give and receive feedback, or avoid the potential for misunderstanding. The purpose of the communication should determine the delivery mechanism.

So carefully consider your goals. Use leaner, text-based media such as email, chat, and bulletin boards when pushing information in one direction—for instance, when circulating routine information and plans, sharing ideas, and collecting simple

Use text-based media such as email when pushing information in one direction. Richer, more interactive tools, like web conferencing, are better suited to problem solving and negotiation.

data. Web conferencing and videoconferencing are richer, more interactive tools better suited to complex tasks such as problem solving and negotiation, which require squaring different ideas and perspectives. Avoid trying to resolve potentially contentious interpersonal issues (telling people that they've made a mistake, that they are not pulling their weight, or that they have upset a teammate) over email or chat; opt instead for richer media to navigate sensitive territory. In short, the more complex the task, the closer you should be to in-person communication. And sometimes meeting face-to-face (if possible) is the best option.

Make intentions clear   Most of our communication these days is text-based. Unfortunately, when text-based tools leave too much to interpretation, common biases and assumptions can cause misunderstandings and lead to unhealthy conflict that hurts team performance. Intentions get lost in translation for reasons such as these:

- **People tend to be less guarded and more negative in writing.** When we cannot see the response of the person receiving the message, it's easier to say things we would not say in person. Emboldened by technology and distance to complain, express anger, or even insult one another, team members can be more negative in writing than they would be face-to-face.
- **Negativity goes both ways.** People on the receiving end of written communication tend to interpret it more negatively than intended by the sender. Emotions are expressed and received mostly through nonverbal cues, which are largely missing from text-based communication. Research suggests that recipients of an email that is intended to convey positive emotions tend to interpret that message as emotionally

neutral. Similarly, an email with a slightly negative tone is likely to be interpreted as more intensely negative than intended.

- **People read with different lenses.** In written messages, we often assume that others will focus on the things we think are important, and we overestimate the extent to which we have made our priorities clear. Unfortunately, it's easy for critical information to get overlooked.

To prevent these biases from causing problems on your team, ensure that you are crystal clear about your intentions. Review important messages before sending them to make sure you have struck the right tone. Err on the side of pumping up the positivity or using emojis to convey emotion and mitigate the tendency toward negative interpretation. Go out of your way to emphasize important information, highlighting parts of the message that require attention, using "response requested" in the subject line, or separating requests into multiple emails to increase the salience of each one.

Stay in sync   When team members don't interact face-to-face, the risk of losing touch and getting out of step is greater. This can happen for a number of reasons. First, when teams are not colocated, it's more difficult to tell when messages have been received and read, unless receipt is specifically acknowledged. Second, communication failures can lead to uneven distribution of information among team members. Individuals might be excluded from an important team email by mistake, for instance, leaving them unwittingly in the dark. Third, the lack of frequent in-person contact can create an out-of-sight, out-of-mind effect in which team members become distracted by local demands and emergencies and forget to keep their distant team-mates informed. When one team member goes silent, the others are left guessing. Without accurate information, people often assume the worst.

Your team can overcome these challenges by making it a priority to keep everyone in the loop. Maintain regular communication with team members, and avoid lengthy silences. Proactively share information about your local situation, including unexpected emergencies, time demands, and priorities. Acknowledge receipt of important messages, even if immediate action isn't possible. And give people the benefit of the doubt. Seek clarification to better understand others' behaviors or intentions before jumping to conclusions. For instance, check in with your teammate who hasn't responded to your time-sensitive message—maybe it hasn't been received, or perhaps something urgent came up.

Be responsive and supportive   The paradox in dispersed team-work is that trust is more critical for effective functioning—but also more difficult to build—than in more traditional teams. Trust between teammates in the same work space is influenced to a large extent by familiarity and liking; however, in dispersed teams, people must signal their trustworthiness by how they work with others on a task. To help develop trust on a virtual team, encourage everyone to respond promptly to requests from their teammates, take the time to provide substantive feedback, proactively suggest solutions to problems the team is facing, and maintain a positive and supportive tone in communications.

Be open and inclusive   Dispersed teams are more likely to have members from different cultures, backgrounds, and experiences.

While diversity can result in a greater variety of ideas, which boosts team creativity and performance, virtual communication sometimes discourages team members from speaking up, making it challenging to capitalize on these benefits. Virtual tools reduce the social cues that help team members bond, which can diminish motivation to share ideas and information. People may also hold back when they can't directly observe teammates' reactions to their contributions. In addition, when dispersed teams consist of subgroups at different locations, there is a natural tendency to communicate more within a local subgroup than across the entire team. This can be particularly challenging for leaders, who may be criticized for unfairly giving more attention to local team members.

To reap the benefits of your virtual team's diversity, focus on communicating as openly and inclusively as possible. Involve the whole team in important communications and decisions. Actively solicit perspectives and viewpoints from all team members, especially those in other locations, to demonstrate openness to different ideas and approaches to a task. And when working to resolve differences of opinion, seek to integrate the best of the team's ideas.

The Role of Leadership

Don't assume that everyone on your team is aware of potential pitfalls with virtual communication or of the five key behaviors that improve performance. We suggest creating a team charter that describes how you will work together. Specify technologies the team will or won't use for different tasks ("Don't use email to discuss sensitive interpersonal issues"); standard formats and etiquette for written communications ("Highlight or bold to emphasize action items in emails"); plans for keeping everyone in sync ("Let the team know ahead of time if a commitment or deadline cannot be met"); expected speed of responses to requests ("Acknowledge receipt within 24 hours"); and types of communication that should always be shared with everyone ("Use the 'would you want to know?' rule of thumb"). We've found that clearly conveyed norms do make a difference.

Our research also shows that people with prior experience in collaborating virtually had higher virtual communication ratings. Leaders can rely on those team members to model effective behaviors—and they can model the behaviors themselves—to raise the whole group to a higher standard.

11

Get Things Done with Smaller Teams

Chris DeBrusk

An important executive goal in most large companies is to improve efficiency and effectiveness. With top-line revenue growth elusive in most markets, a key way to increase returns to shareholders is to boost the bottom line—and that means stepping up productivity. These gains need to come from improving the processes that run the company as well as those that change it.

Unfortunately, achieving greater productivity in project teams focused on change can be challenging, especially when new technology is involved. Every company has experienced a project that was either delivered at twice the budget and in double the time, or never actually delivered against its objectives and eventually scrapped. There are many reasons why these large programs fail, but one potential root cause is that they simply break down under their own weight. One way to improve the effectiveness of projects is to reduce the size of the teams mobilized to tackle them. In other words, it might be time to make your project teams smaller.

Smaller teams move faster, iterate at a higher frequency, and innovate more for the company. There are endless examples of small teams achieving amazing things. When Facebook purchased WhatsApp for $19 billion, the company's 32 engineers had created a platform that was used by 450 million users. The Volkswagen Golf GTI, one of the most famous hot hatchbacks in history, was created by a team of eight. Many of the largest technology companies created their first successful products with teams of fewer than 10 people.

Jeff Bezos famously instituted a "two-pizza rule" in the early days of Amazon. His edict was that any team that could not be fed by two pizzas was too big. In concept, it is fairly easy to understand how a smaller team can be more effective, as communication is easier and decision making can be accomplished more quickly.

But practically, how can managers take advantage of this technique in large organizations?

Make Big Problems Smaller Problems

Establishing two-pizza teams is especially challenging with in-flight programs in large organizations that tend to grow over time. If you track the scope of a major change initiative, you'll often find that by the end of the project, the goals of the program bear little resemblance to those that were agreed on at the beginning. This growth effect is vastly multiplied as the size of the program and the team supporting it grows. New team members bring new goals that have to be incorporated into the program (and, of course, nothing is ever removed).

One way to control this natural program weight gain is to break down the project from the beginning into discrete problems that can be solved by smaller teams. With leadership focus, even large, complex problems can be compartmentalized into separate, achievable pieces that a small team can easily take on and over deliver.

Another way to get smaller without actually shrinking your organization is to break business capabilities into focused organizational units, each with a clear mandate to provide the company with a discrete set of services that are well defined and understood. Then put someone in charge and give them responsibility and authority to get things done—and see what they can accomplish.

Ensure No Team Member Is Indispensable

One of the realities of many large organizations is that few people have the luxury of working only on a single problem. As problems and the teams that aim to solve them multiply, key people get split across many competing tasks, so their time is sliced into small, difficult-to-manage increments. When a critical member of a team is simultaneously working as a member of six other teams, delivery against critical milestones is inevitably affected. Just getting on that person's calendar becomes difficult.

It is important to actively manage away from "indispensable" people in your change and transformation projects. Single-threading important decisions to any one person creates a single point of failure for the team. As a result, projects and innovations move at a slower pace and operate with higher risk. Small teams allow you to take steps more easily to cross-train as a way to manage this risk.

Adopt One-Step Decisions

Large teams are notorious for needing multiple steps to make most decisions. Aligning calendars often takes time, and once you get everyone into a room (or, more likely, on a call), several attendees need to be brought up to speed. Some attendees will not have read the requisite material, and others will have been sent as substitutes for key decision-makers who could not make the time (and these substitutes will not be able to make any critical decisions without conferring with their boss). We've all attended these sorts of meetings. They rarely result in decisions— and they usually lead to additional meetings. A small team can shortcut these issues much more easily. Fewer people need to be present to make decisions, and those present are typically much more involved in the details of the problem, so they don't need a meeting to ramp up before they can contribute.

When you first form a team, spend time to determine what types of decisions each member of the team can make on their own, what types of decisions the team can make as a group, and what needs to be escalated for more senior input. Then fight hard to push as much of the decision making to the team by defining very clear guidelines that give the team ownership and accountability.

Build Trust

There is nothing more powerful in a team than trust. It accelerates progress, improves quality, and reduces execution risk. Yet trust doesn't come automatically and often needs to be intentionally created. Smaller teams allow managers to spend more time with each person, getting to know their strengths, weaknesses, and career goals. They can structure tasks in a way that reinforces the natural strengths of the team, which allows team members to show their competence, and that builds trust. Team members also build trust through constant interactions as they tackle and solve problems together.

A successful team event outside of work doesn't hurt either. Sometimes trust is built in the bowling alley, the paintball pitch, the basketball court, or just enjoying good food.

Be Less Formal When Sharing Information

Presentations are an amazing tool for communicating complex ideas. They are also a huge time sink for teams. By moving to smaller, more focused teams, you can reduce, if not eliminate, the need for the structured communication that a presentation provides.

A small, focused team can easily replace a formal presentation and slide deck that requires hours to create with a whiteboard session where the problem and solution are diagrammed out in real time. If the team is not colocated, there are many collaboration tools like Slack, Microsoft Teams, and Symphony that can replace a physical whiteboard and move the brainstorming online. The impact is the same: faster solutions and greater alignment across the team.

Increase Visibility (and Accountability)

It is fairly easy to hide in a large team. You can dial into conference calls and avoid contributing. Since deliverables are owned by multiple people, it is easy to let others do the heavy lifting. You can be busy during key meetings where you need to be prepared. We've all worked on teams in which no one really knew what some members were doing.

With small teams, hiding is nearly impossible. A lack of contribution is immediately noticed, and those who don't contribute to moving the ball forward can be moved off the team much easier because there is direct evidence that they are not adding value.

Limit Unnecessary Synchronous Meetings

While it clearly isn't possible to eliminate conference calls entirely, the ubiquity of collaboration platforms like Slack, instant messaging, and desktop videoconferencing means that teams can communicate in many ways that don't revolve around conference calls with all of their inherent challenges.

Small teams can avoid conference calls because they communicate often, through multiple channels, both digital and face-to-face. Large teams don't interact nearly as much and therefore need regular catch-ups to re-sync on objectives and ensure information is being shared effectively. Even then, conference calls are still relatively ineffective in ensuring alignment.

Focus Less on Tracking Project Progress

Project managers live to track things. They make lists of tasks and track progress against them. They identify dependencies and track relative slippage. An effective project manager is able to act like oil for a large project team, reducing friction just when it threatens to affect delivery. Yet less effective project managers spend their days bothering teams for status updates and creating presentations aggregating the updates they received.

With the widespread adoption of tools like Slack and Jira, the need for someone to play the role of progress aggregator is quickly coming to an end because the platforms automate tracking. This is even more pronounced with smaller teams, who communicate at such high bandwidth that everyone knows what tasks are behind and what needs to be done to get them back on track.

Product managers are critical, but the usefulness of project managers should be questioned, especially as teams get smaller and more focused.

Work More Easily with Other Teams

As you break down programs into chunks that smaller teams can tackle, the interface between teams dependent on each other becomes critical. It is often friction between teams that turns into missed dependencies and timeline slippage.

It is important to craft "contracts" between teams that clearly outline their individual mandates, spell out how they will interact, and identify what they can rely on each other to provide and when. A focus on clear and concise dependency management increases transparency while ensuring that each team can focus on achieving the specific goals it has been assigned.

Embrace Technology Faster and More Effectively

Smaller, empowered business teams that have control of their own destiny start to figure out how to leverage technology to improve the world they now control more quickly—especially if you give them more direct oversight of the technology teams that support them.

In order to accomplish this, companies need to figure out which aspects of their technology delivery capability need to be centralized and which aspects can be decentralized and moved closer to the teams who are executing for customers and colleagues. Improvements also need to be made in the process of evaluating and approving the use of innovative technologies, as this is often a point of friction in larger organizations.

The emergence of machine learning as a critical business tool is an example that shows how technology teams are working to put sophisticated analytics capability into the hands of small teams of business users. While difficult to accomplish, a properly balanced technology delivery capability that supports execution teams will result in increased innovation and acceleration in meeting business goals.

As technology gets more modular and flexible and as organizations adopt agile delivery techniques, the concepts outlined above are likely to become more mainstream. If you want to reduce execution risk, increase the pace of innovation, and deliver faster, turn your big project into a group of smaller projects and let the teams get to work.

12

Is HR Missing the Point on Performance Feedback?

Sergey Gorbatov and Angela Lane

The field of performance management has been in turmoil lately. Employees are getting confused. Leaders are getting frustrated. Consultants are getting rich. Why the upheaval? A small number of high-profile companies (including Adobe, GE, and Accenture)[1] have abandoned traditional performance management in favor of less formal and quantifiable approaches that prioritize ongoing conversation over annual ratings. Elsewhere, human resources executives have dashed to follow suit, assuming what is best for these industry leaders will be best for their organizations, too. The problem is that there is little evidence to support these new approaches. We are not specifically arguing in favor of traditional ratings, nor do we believe most legacy performance and feedback systems are built to address today's talent management challenges. But we are arguing for a pause before jumping into a hasty performance management redesign.

Human performance is complex. Some may even call it messy. Yet there are rules, principles, and science behind performance management. While the ultimate goal of any HR initiative is to improve performance, numerous intermediary levers can affect the outcome. Feedback is one of them, but there are many others, including goal setting, context, deliberate practice, and rewards structure. The answer to complexity is not oversimplification; the right process is the one that leverages the known facts about what drives performance. Today, HR too often ignores these facts, instead chasing "bright, shiny objects," what John Boudreau and Steven Rice have referred to as flavor-of-the-day HR novelties.[2]

HR practitioners are too quick to remove a process that corporate leaders don't like. They should be arguing for the tough improvements that would actually drive performance, beginning with the use of feedback.

Myths about Performance

What are some of the myths that have led HR to redesign processes, despite the science? And what are the implications for a quality performance management process?

Myth No. 1: It's a problem that employees don't like formal performance feedback   Yes, there is certainly evidence that people don't like feedback. Employees dislike it to such an extent that many will dodge feedback opportunities if they can.[3] But that is hardly a reason for managers not to provide it. What an employee "likes" and whether he or she is *satisfied* are different. An employee's *satisfaction with feedback* has a powerful mediating impact on performance; if he or she is dissatisfied with feedback, other performance factors, such as accountability or confidence in performing the task, plummet.[4] Thus, the best system is not one that reduces the aspects of review, like a final rating, that employees or managers "don't like." The best system is one that delivers feedback that *satisfies*. If a rating summarizes feedback that is clear, fair, developmental, sensitive to employees' needs, and more, everyone will likely be satisfied.

Consider an annual checkup with your dentist. You may not like the conclusion that you need a filling. But you can be satisfied that the diagnosis is thorough, and the message can be delivered without making you feel like a jerk for not brushing or flossing adequately. And regardless of how "happy" you are, addressing the problem is the best possible outcome.

Myth No. 2: "Bad" feedback is . . . bad   The belief that negative feedback is bad is reflected in performance management redesign, which focuses on "appreciation," for fear of demotivating the employee. The risk of this approach is that leaders avoid giving feedback, or sugarcoat it to the point that it has no practical value, or simply deliver it poorly. Indeed, leaders are notoriously unskilled as feedback messengers: Global data shows that the skill of giving constructive feedback is at the bottom of the competency list for managers and executives.[5] Perhaps, then, it is no surprise that whether feedback is positive or negative turns out to have virtually no bearing on performance. Studies show that positive feedback may lead to a decrease in effort, just as negative feedback may boost one's desire to achieve more. Also, receiving only positive feedback keeps people from taking in negative feedback long term. An experimental study of strategic decision-makers showed that resting on the laurels of past successes and consistently getting only positive feedback will impede people from listening to negative feedback in the future, when it may actually be needed to correct a faulty course of action.[6]

When feedback is properly situational, it loses its bad rap. Several factors should be taken into consideration to match the right message with the right person in a concrete situation. For instance, motivation improves both when positive feedback is given to people with a promotion focus (those who want to achieve and take risk, who are sensitive to rewards) and when negative feedback is given to people with a prevention focus (those who want to avoid trouble and are generally cautious, who are sensitive to punishment).[7] Negative feedback can be especially beneficial in "critical events"—novel, uncertain, first-time situations when the individual cares deeply about the outcome, such as leading a new team or managing a crisis.[8] Yet a blanket belief that bad feedback is bad aligns with most people's desire to avoid conflict. It is easier to believe that feedback needs to be motivational and uplifting—a clearly mistaken assumption. Rather, leaders must adopt the approach that is best aligned to the business challenges of the organization and the individuals involved.

Myth No. 3: If feedback is good, then frequent feedback is better Much of the drive toward removing annual performance cycles in favor of regular check-ins has its origins in this sound bite. This is a case of "Yes, but . . ." While there is evidence that frequent feedback is good, feedback should not be excessively frequent. A study at a large Midwestern university demonstrated that there is a tipping point at which an increase in feedback frequency leads to a decrease in task effort and performance. This goes against the conventional wisdom that employees need a lot of feedback, especially as they are learning a new task or a new role. A study conducted a few years ago tells us that too much feedback can be particularly harmful at the early learning stage.[9] Of course, this is only one study. Yet it's a helpful reminder that not all our long-held assumptions may be correct—and our mistaken assumptions may come back to haunt us if we hinge business decisions on them.

Here we also see a false dichotomy: If I want frequent feedback, I must dispense with an annual performance cycle. And that drives down the quality of the manager-employee conversations.[10] Instead of thinking expansively, in terms of "Yes, and . . . ," HR responds by removing existing processes. There are notable examples of companies that take a comprehensive view of performance management and the role of feedback. Facebook is one of them. Refusing to board the abandon-the-ratings bandwagon, the company published the article "Let's Not Kill Performance Evaluations Yet."[11] Frequent feedback and regular performance evaluations may be complementary, not mutually exclusive.

Myth No. 4: Managers are essential to the performance management process Most HR leaders would accept this as fact. Yet if there is lack of trust in the manager-employee relationship, the weight of the feedback decreases dramatically. We learned from a study of bank workers that source credibility—trust in the person giving feedback—strongly correlates with perceived accuracy and with a desire to respond, both of which have an impact on performance.[12] When trust and engagement with managers are low, feedback won't drive the desired outcomes. HR must then look for alternative sources of feedback, such as a 360-degree assessment, because feedback from a low-credibility leader won't change anything.

Myth No. 5: Performance gets better with feedback This is another case of "Yes, but . . ." A careful review of scientific data affirms that simply providing feedback does not necessarily move the needle. One of the most influential meta-analytical studies on feedback demonstrated that only about a third of feedback interventions result in improved performance.[13] That is, the remaining two-thirds do not. For this myth to become truth, the manager has to deliver quality feedback that takes into account a variety of factors (such as task, context, and personal characteristics) and synthesizes them into an appropriate message. Effective messages may compare one's performance with the previous year's, with peers', or with best exemplars in the organization. Such a message could, in fact, be a performance rating. When people don't know how they objectively compare with others, they are less likely to put in discretionary effort.[14] A rating is a helpful mechanism to ensure consistency in delivering the organizational message to the individual about the level of performance relative to expectations.

Myth No. 6: Feedback just happens   This magical thinking is refuted in study after study. Feedback-rich cultures do not appear out of thin air but depend on structure, processes, and persistence. Even those managers who indicate they know what to say to help employees develop, and how to say it, need a support process to guide them. Out of 500 managers surveyed globally, about a third confessed that they did not know how to help people change, and less than 10% said they knew how to make such behavioral change sustainable.[15] A change to the process that results in either poorer-quality feedback or inadvertently less feedback will likely cause a decline in performance. For example, CEB Inc. discovered that fewer hours are spent in informal performance conversations in organizations without performance ratings, compared with those that give ratings.[16] The same study revealed that employee perception of the quality of the feedback conversation is also 14% lower in organizations without ratings. Oversimplifying performance management in times when the managers need structure most is hardly responsible.

Changes to performance management that respond to just one or two chosen facts, while ignoring the complexity of human performance, waste an opportunity to drive individual performance and organizational success.

The examples above indicate the complexity of performance management and the dangers of relying on buzzwords and sound bites to redesign processes. We believe HR has a responsibility to understand this complexity and provide clear and concise guidance to managers on how to give feedback and manage performance in ways that drive business results. This requires an appreciation of why feedback matters and what makes feedback "good."

Why Feedback Does Matter

What do we really know about what happens with feedback in organizations?

We know that successful leaders know themselves. We know that feedback is a key component in gaining this self-awareness, and self-aware people are more successful. In their seminal book *The Leadership Machine*, Mike Lombardo and Bob Eichinger outline six key sources of personal growth, and feedback is one of them.[17] We know that feedback can move the needle, and that appropriately frequent and repeated feedback spurs personal change to take place.

When the feedback process is well managed, meaning it is perceived as credible and accurate and is received in the manner it was intended, it has a significant positive correlation with performance.[18] Of course, other individual and organizational aspects contribute, such as a feedback-rich environment and the individual's desire to engage in the conversation, and we must embrace this multidimensionality. Improving feedback is not only about training managers, but also about organizational culture, a sharp focus on performance, and holding leaders accountable for people development. Still, the impact of feedback on the contribution a manager makes to the organization is hard to ignore.

Understanding what good feedback looks like, and how it can be delivered, will help win over even the most feedback-averse manager. That is the first step. The second is designing a solution to take performance to the next level.

Making Feedback a Good Investment

We think of feedback as an investment of time and resources: Bad investments destroy value, good investments add value, and the best investments create sustainable value. How does HR make feedback a "best investment"?

Create an approach that links feedback to the business strategy or company philosophy Increasing value is the main driver of all actions in business environments. When giving feedback, take time to explain how the business is doing. Show how individual goals relate to bigger enterprise objectives and strategies. Highlight the connections between the individual's work and business needs. And importantly, explain why the subject matter of the feedback you are about to give is important to the business.

Contrary to the view that too much information overwhelms the employee, providing a multifaceted and comprehensive description of the business situation, combined with a clear message on what is important, has a positive impact on performance. Data from U.S. Air Force research proves that employees can use complex, nonlinear information when it is included in feedback.[19] And what are better words to describe the current business environment than "complex" and "nonlinear"?

Ensure your system focuses on performance first, development second Focusing on performance goals gets people to act more quickly and with more zeal than a personal development target does.[20] It also diverts the focus from the individual to the task, reducing the potential for negative reactions to feedback. Indeed, data from experiments at the University of California, Los Angeles, suggests that when information is ego-relevant—directly affecting one's opinion of oneself—the feedback is more likely to be misinterpreted.[21]

How performance objectives are framed, though, can make a big difference. While the manager's main goal is to increase performance, the positioning of feedback works best when it contains the expectation of change and when it provokes the recipient to think in terms of a learning goal, such as, "Yes, I am keen to know how to design responsive web pages." In one study, a group of students were given two challenging tasks separated by a period of time. They received feedback on how well they performed the first time around. Those with a learning-goal orientation performed better on the second task. Those with a proving-goal orientation (for example, "I must prove I am worthy of a promotion" or "I must do better than George on this") did worse.[22]

Employee satisfaction with feedback creates a virtuous circle. It has a powerful and additive impact on performance. When employees receive feedback that they perceive to be valuable and intended to help them improve, other factors of performance are unlocked: utility ("I believe feedback will help me achieve my goals"), accountability ("I ought to do my job well"), self-efficacy ("I am able to perform the task"), and social awareness ("I take others' opinions of my work seriously").[23]

Ensure leaders focus on both the what *and* the how   In many companies, the way in which employees get results has a great impact on their overall performance evaluations. Getting stuff done is important. But feedback on *how* they get results can change outcomes. Giving employees feedback on how their behaviors help or hamper their ability to achieve the desired business outcomes will be more effective and position them for success in the future.[24] Managers should provide guidance on both productive and destructive behaviors. The nature of "derailers"—behaviors that impede success—is such that most people are either unaware of theirs or unable to control them without help. Focusing only on strengths can be detrimental to individuals' careers, as organizations are quite unforgiving about limitations—managers deselect for things that people are not good at.[25] Delivering performance feedback in a manner that helps the employee realize his or her own nonproductive behaviors leads to a better mood at work, greater job satisfaction, and stronger organizational commitment.[26]

Ensure that feedback is both specific and generic   A study was conducted with two groups of students taking part in a computer simulation. One group received generic feedback on their people management decisions, and the other got more specific feedback. The group receiving detailed feedback did better on the task, but the group getting generic feedback had to think more and therefore demonstrated a greater degree of competence after the simulation. Specificity increases task performance, while broad guidance promotes reflection and enhances learning.[27] Clearly, both organizations and employees will benefit most from a combination of the two. For specificity, point out the exact behaviors that need to be improved, using facts and concrete situations from the recent past, and relate them to customer or peer feedback, if available. To switch to the more generic level, ask open-ended questions ("Changing which one behavior will make the biggest difference for your results?"), link individual results with the company's strategy, or encourage employees to compare their performance with that of the best in the company and the industry.

Aim for structure, consistency, frequency, and immediacy
There are moments in life when less is more. Choose one or two critical points to deliver and focus on those. Giving feedback on every aspect of how employees go about their work will cause confusion and stress. Being structured and focused helps.

Similarly, consistency from one feedback interaction to another reinforces the message and reminds the employee to stay on track. The same message, delivered consistently, frequently, and via a variety of means, builds a solid foundation for behavioral change.

As mentioned earlier, feedback delivered too often may reduce both learning and task performance,[28] but the consensus is that immediacy and frequency are desirable.[29]

A well-designed performance management system promotes frequent and immediate feedback. It encourages feedback that is structured, and drives consistency both vertically within a department or a division and horizontally across the organization. It makes it easy for leaders to identify and prioritize behaviors, by restricting feedback to critical points or limiting feedback to selected behaviors.

Add context and perspective   When assessing performance, managers should take into account information from multiple sources, both internal and external. They must consider the complexity of the operating environment, the adversity of situations, and the support that has or has not been provided to help employees attain their goals. We call these factors "feedback modifiers." They complicate the job of delivering accurate and useful feedback, but they are essential to how the employee will see the credibility of the feedback provider and to the accuracy of the feedback. The fact that this adds complexity is why those who lead others are paid a premium.

Remove "noise"   Managerial discretion is part of performance evaluation and, we believe, usually a positive aspect. We know that managers' assessments are typically more accurate than self-assessments.[30] Managerial discretion also enables leaders to take account of everything that determines performance, including mitigating factors that would rightly impact perceptions of fairness. However, this same discretion can have negative consequences—through managers' unconscious biases or more malevolent factors, such as the politics of the organization.[31]

These noise factors can make it hard to provide fair, consistent, and useful feedback. HR should be accountable to minimize the noise. Examples of noise include:

- **Lack of transparency.** Establishing clear rules about what, how, and when feedback should be shared will mandate fairer and more open conversations.

- **Organizational culture.** The culture will either help or hinder the difficult task of giving fair feedback. If the culture is less open, HR must design more structured and formal processes. If the culture is open and candid, less structured processes can be implemented. Changes based on external trends, which do not take account of the prevailing company culture, risk diminishing the value of feedback and, ultimately, performance.

- **Biases.** HR has a role to ensure that leaders are educated on bias and that there are clear expectations that leaders actively work to minimize biases. Systems of calibration help here, as does the prospect of third-party review of assessments. We have written separately about ways to calibrate feedback using sources such as comparisons with internal and external peers and viewpoints of stakeholders and clients, and having managers check in with their own boss.[32] Also, knowing that someone else will be auditing the feedback later will keep managers' biases more in check and deter political maneuvering. While it is impossible to eliminate bias, leader awareness and conscious checking against the most pertinent forms of social (gender, age, race) and cognitive bias in performance appraisals will enhance the quality and outcome of feedback. Cognitive bias is less visible than the other kinds and may seem innocuous, but it has a real capacity to cripple decisions. Such bias comes in many shapes and forms—for example, the "halo and horns" effect (in which singular positive or negative experiences color an overall assessment), confirmation bias (seeking out and giving weight to information consistent with one's own views while discarding discrepant data), or affinity bias (having a more favorable impression of someone who is similar to oneself in some way). Making people aware of their biases, training, and just-in-time priming (providing reminders to be aware of their own tendencies to succumb to bias) can all help managers make better decisions.

When managers have the right understanding of feedback, their job becomes much easier. We change and update our views about how the world works all the time. A danger of that is that we are likely to be drawn to bright, shiny objects or "trends du jour"; psychologists call these *salience* and *recency* effects. HR must help organizations and their leaders guard against such cognitive biases and work hard on building beliefs based on scientific evidence.

GE's Jack Welch once said, "As a manager, you owe candor to your people,"[33] and collectively we need to do a better job at enabling better, more candid feedback. But, of course, it takes two to tango, and it may be unfair to lay blame on managers alone. Employees engage in all types of feedback-avoiding behaviors to preserve positive impressions in the eyes of others, to avoid appearing weak or incompetent, and to maintain self-esteem.[34] They may be recalcitrant, only pretend to care, or genuinely not "hear" the message. All of these, too, are the manager's job to address by—surprise!—giving timely and accurate feedback.

We conclude that a lack of honest, focused, and timely performance feedback is a major constraint to organizational performance. Unfortunately, we see too many examples of HR ignoring the difficult facts in favor of simplistic, flavor-of-the-month solutions that will only make matters worse. Lombardo and Eichinger, experts on derailment, put it in no uncertain terms: "Getting no developmental feedback or feedback on strengths alone are time bombs which explode in managerial and executive roles."[35] It is HR's job to take on the difficult challenge of instilling a robust and comprehensive approach to feedback and performance management that fits the particular needs of the organization. The approach must give managers the tools to truly improve a company's performance and give employees the path to self- and career improvement.

This article represents the authors' personal opinions and not those of their employers.

13

The Leadership Demands of Extreme Teaming

Amy Edmondson, interviewed by Frieda Klotz

Technology has made business more globally connected than ever before, allowing organizations to join forces across professions, geographies, and industries. This is especially true for innovation projects, where diverse experts bring their specialized knowledge into play.

But there's a hitch: These kinds of team projects have built-in hurdles because of differing communication styles, cultures, and professional norms.

Amy Edmondson says many managers are not equipped with the skills to capture the full value of these multifaceted collaborations. Edmondson is the Novartis Professor of Leadership and Management at Harvard Business School and coauthor, with Jean-François Harvey, assistant professor at HEC Montréal, of *Extreme Teaming: Lessons in Complex, Cross-Sector Leadership* (Emerald Publishing, 2017). Learning how to navigate these new challenges is crucial, Edmondson says. She predicts that a more active concept of "teaming" will gradually replace the notion of teams, with increasing numbers of fluid, temporary assignments that cross multiple boundaries.

MIT Sloan Management Review spoke with Edmondson about these complex collaborations and the skills needed to manage them. *MIT SMR* correspondent Frieda Klotz conducted the interview, and what follows is an edited and condensed version of their conversation.

MIT Sloan Management Review: Can you please define "extreme teaming" and explain how it's different from what most of us think of as teams?

Amy Edmondson: A team is a bounded, interdependent group of people responsible for a shared outcome. However, with 24/7 global operations, complicated shift patterns, and changing tasks and work needs, more workplaces today require people to collaborate to get things done outside of the context of a formal team.

I have been using the term "teaming" for a while to capture the fact that more people are finding themselves having to collaborate across boundaries without the luxury of a stable team structure. Many of those boundaries are across distance, but many are also across disciplinary expertise or hierarchies of power and status. Extreme teaming is a term that Jean-François Harvey, my coauthor, who teaches at HEC Montréal, and I came up with. It captures not just teaming across functions or time zones for people working in the same company, but teaming that extends across organizational boundaries and sometimes even industry boundaries, since many innovation challenges call upon people to work with people from other organizations.

This form of teaming was interesting to me because I'm a social psychologist at heart—my training is in organizational behavior and social psychology. When I think about human beings and the interpersonal dynamics between people, and then about these new opportunities to team across sectors, I say, "Wow, that's not going to be easy. That's going to take some new skills, some new mindsets, and some new thinking."

Why do so many teams fall into this category of extreme teaming, in your view?

Beyond the increasingly globalized workplaces, there's a recognition that we're not always going to rely on vertical integration to solve all of our challenges. It sometimes just makes sense to team up with another organization to get something done.

For example, you might be a hospital working with a software company to design a new system for monitoring patient safety, but you don't hire all those people, you just work with them. That makes good sense. At the same time, it takes a while for people to get up to speed and learn one another's professional languages. Sometimes they don't have that time because people are [constantly] shifting in and out of the group.

You mention in your book that experts are increasingly specialized. Presumably that plays a part in extreme teaming?

Right, that's an important part. With the knowledge explosion, we get more and more specialization, which also implies narrower and narrower specialization. Most of the innovation challenges we're talking about in the book are not solved within a single narrow area of expertise but require people to work across expertise boundaries. That can be hard because we don't always understand one another's expertise or even one another's outlook.

Can you give us an example of extreme teaming in action?

The Chilean mine rescue—[the operation] to evacuate 33 people who became trapped underground in 2010—led to a huge, 69-

day, cross-sector collaboration involving the mining industry, experts from other industries, and [experts] from the military and government sectors. NASA was involved, the logistics company UPS donated air transport of specialized equipment. Many groups came together and teamed up about as well as you can imagine, under pressure and in a crisis situation.

Some people say the fact that it was a crisis made teamwork easier. That may be true, yet there are many times when we see crises in which people don't come together extraordinarily well. I would attribute what happened in Chile to leadership—leadership at multiple levels, including underground, at the top of the country, and at the top of the rescue operation, which was an innovation project in the truest sense. There was no solution at the outset. By teaming up across national and expertise boundaries, the group collaboratively developed novel solutions.

What skills do people need to work in this way? Are they largely communication skills?

Communication skills cover a lot of territory, actually. So, yes, in a deep way—communication skills, including empathy and curiosity, are crucial. Leaders can be skilled at articulating their thoughts or skilled at listening, but neither is enough.

Leaders must also have a high level of self-awareness to keep reminding themselves of the things that they are missing, because each of us is under the illusion that we see "reality," or that our perspective is a good map of reality.

So there's humility involved, then?

Yes. Curiosity, empathy, and humility are three qualities that I often come back to. Not a false humility, but a genuine, situational humility—"We've never been in this situation before, so

Most managers
have been either
explicitly or
implicitly trained
to think in terms
of accomplishing
fixed goals, tasks,
and deliverables in a
predictable world.

I'm confident that I don't know everything. I have to remind myself to be fully aware of that."

What's the role of technology in all this?

Teaming would not be possible without technology. Imagine if you recognized your need to work with someone in another organization or location or part of the world but had no access to information technology. You simply wouldn't be able to do it without the technology to facilitate it. Technology is often imperfect and frustrating, but it's vital and it starts the ball rolling.

In your book, you say that teams are "the performance unit par excellence for innovation." Can you talk about the sorts of innovation that extreme teaming helps bring about?

The majority of innovation projects can be carried out within the four walls of the organization. But for instance, for projects where government permitting is involved, an organization needs to work with city hall, but of course that doesn't mean employing city hall. The kinds of projects that are inherently multisectoral bring up additional challenges like professional culture clashes and a need to navigate different, taken-for-granted time frames and professional norms.

What particular industries lend themselves to this sort of teamwork?

I have done a lot of research in the context of health care delivery. The challenges that health care faces are immense. There's a fundamental shift under way from fee-for-service medicine to being paid for value, which means a fundamental shift toward focusing on health rather than just on sick care—which, of course, clinicians will continue to care about.

The shift requires health care practitioners to think about ways to help people stay healthy, which is outside what they have focused on in the past. In the future, simply providing more care will not guarantee more income, and finding ways to provide value in the form of the health of a given population will be vital to the success of the health care industry. In many cases, this will mean that health care providers will be teaming up with people in communities with different skills and responsibilities. This might mean partnering with community organizations, with schools, and with companies to promote a culture of health in ways they've never had to do before.

Any industry confronting large trends with implications for how work is done is an industry that's ripe for new thinking and for these sorts of cross-sector collaborations.

You write that "most managers remain ill equipped to effectively lead extreme teaming endeavors because these collaborations pose different challenges than those that managers typically face when leading teams inside their organizations." How can managers be more successful?

When I say that managers are ill-equipped, I mean that most managers have been either explicitly or implicitly trained to think in terms of accomplishing fixed goals, tasks, and deliverables in a predictable world. We all know we're not in that kind of world—and yet the fundamental mindset and skills of management work best for fixed, understandable, reasonably predictable deliverables.

We can all learn to be curious, empathetic, humble, and deeply interested in someone else's perspective. But it's not a given. It requires an adjustment to say, "I don't have the answers" or "Management is about generating reasonable hypotheses from

We can all learn
to be curious,
empathetic,
humble, and
deeply interested
in someone else's
perspective. But
it's not a given.

what we know at any given point in time." In that sense, management is a lot like science. [As with] science, you can view all actions as tests of those hypotheses and as opportunities for collecting data. You'll figure out what works, what doesn't, and what to do differently the next time.

Is less top-down management required?

I think so. It almost no longer makes sense to think that I, as the manager, could be best positioned to fully evaluate someone else as the subordinate. They see things I miss, I see things they miss, so it's got to become more of a conversation. Managers are more like coaches. Oftentimes managers have a slightly better perspective because of where they sit, but they don't have omniscience. Few managers should see themselves anymore as the boss in the traditional sense or as the person more likely to be right compared to a subordinate.

That's interesting. I'm sure some managers would find it difficult.

Yes. And yet we all have to become coaches and direction-setters—but ones who are completely open to a range of possibilities.

14

If You Cut Employees Some Slack, Will They Innovate?

Yasser Rahrovani, Alain Pinsonneault, and Robert D. Austin

The idea of using slack resources—in the form of time, technology, and support—to bolster employee innovation falls in and out of favor. The return on slack innovation programs can be prodigious: 3M attributes the development of the Post-it Note to its 1948 decision to allow employees to devote 15% of their paid time to side projects; and Google says its "20% rule," which upped the ante on slack time devoted to innovation, yielded Gmail, AdSense, and Google Earth. But few, if any companies, have stuck with time off for innovation and other slack-based programs for as long as 3M. Even Google has reportedly waxed and waned in its commitment to its 20% rule.[1]

Given the significant investment that slack-based innovation programs require, the decision to adopt one shouldn't be made off the cuff. But what are the factors underlying that decision, and how should such programs be designed? To begin to answer these questions, we conducted in-depth interviews of knowledge workers in different industries to understand what motivated them to take risks and explore new ideas and, more specifically, whether and how slack resources might have contributed to their innovativeness. We then created and refined an empirical model based on the factors and relationships that appear to influence employee innovation and tested it using a sample group consisting of 427 employees from North American companies.

We found that different types of employees respond in different ways to slack innovation programs; that different kinds of slack resources are better suited to certain types of employees than they are to others; and that different kinds of slack innovation programs produce different kinds of innovation. Companies can use these findings to design more effective slack innovation programs and maximize their returns on slack resources.

A Tale of Four Employees

Every employee is unique, but for the purposes of this research, we focused on two employee dimensions that are particularly relevant to innovation: level of job expertise and how innovative people consider themselves to be. This yielded four types of employees who form the boundaries of the workforce at large.

- **High expertise, high innovation (HEHI):** Employees who rank high in job-related expertise and in their personal assessments of their own innovativeness. They are curious and love to learn, particularly about technology, and they seek out new technologies.
- **High expertise, low innovation (HELI):** Employees who rank high in job-related expertise and low in self-assessed innovation. They take refuge in their expertise, prefer stability in job processes, and dislike the idea of shifting to new technologies.
- **Low expertise, high innovation (LEHI):** Employees who don't have high levels of expertise but are eager to learn and try new technologies. Often, they are new to their jobs.

- **Low expertise, low innovation (LELI):** Employees who don't have high levels of expertise and aren't comfortable with innovation and change.

These four employee types represent the extremes of our sample. Pure HEHI and LELI types represent only about 8% of employees each, and pure HELI and LEHI types represent about 1% each. The vast majority of the people in our sample group (359 of 427) fall somewhere in the middle: They are within a standard deviation of the mean on both expertise and self-assessed innovativeness. We believe that this is the norm for the typical workforce: lots of people in the middle, a few especially strong, a few weak (or new), and not very many "off-diagonal" oddballs (extremely high in one attribute, extremely low in the other).

Of course, workforces vary, too. We expect a company like Google—an innovative organization in a technically complex industry—to skew toward the HEHI types while more conventional enterprises like oil companies or traditional carmakers would likely have more typical distributions of employees. Why does this matter? Because managing innovation (and the design of slack innovation programs) in a workforce that is disproportionately made up of HEHIs and people on their way to becoming HEHIs might be quite a bit different from managing innovation in a more typical workforce.

Different Strokes for Different Folks

Perhaps the broadest conclusion that emerged from our research is that HEHI employees are different from the other three types of employees in terms of the management levers that encourage them to innovate.[2] Motivational science helps explain why.

There are two different kinds of motivation that concern us: intrinsic motivation and social motivation. Intrinsic motivation is oriented primarily toward self-development, pursuit of one's own interests, or self-fulfillment.[3] Social motivation is oriented primarily externally, toward helping others.[4] When motivated socially, employees may improve their own work practices, but

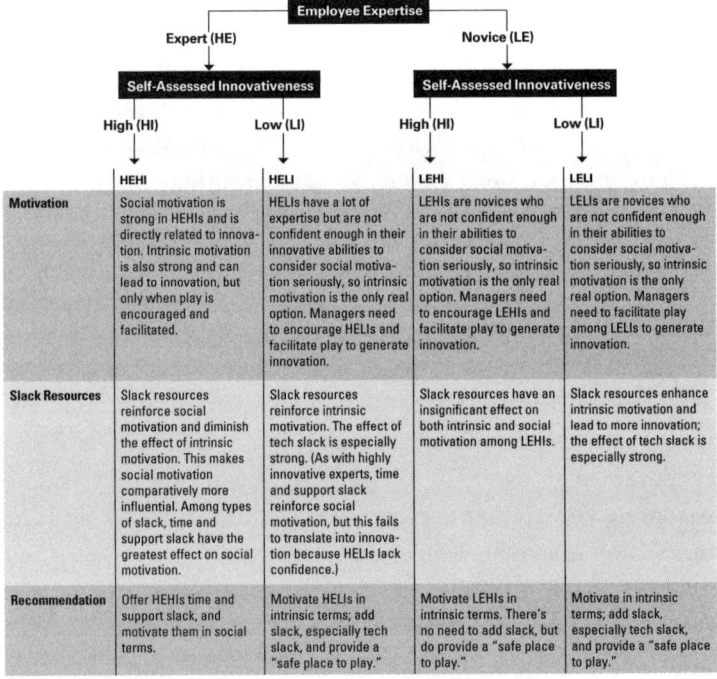

	HEHI	HELI	LEHI	LELI
Motivation	Social motivation is strong in HEHIs and is directly related to innovation. Intrinsic motivation is also strong and can lead to innovation, but only when play is encouraged and facilitated.	HELIs have a lot of expertise but are not confident enough in their innovative abilities to consider social motivation seriously, so intrinsic motivation is the only real option. Managers need to encourage HELIs and facilitate play to generate innovation.	LEHIs are novices who are not confident enough in their abilities to consider social motivation seriously, so intrinsic motivation is the only real option. Managers need to encourage LEHIs and facilitate play to generate innovation.	LELIs are novices who are not confident enough in their abilities to consider social motivation seriously, so intrinsic motivation is the only real option. Managers need to facilitate play among LELIs to generate innovation.
Slack Resources	Slack resources reinforce social motivation and diminish the effect of intrinsic motivation. This makes social motivation comparatively more influential. Among types of slack, time and support slack have the greatest effect on social motivation.	Slack resources reinforce intrinsic motivation. The effect of tech slack is especially strong. (As with highly innovative experts, time and support slack reinforce social motivation, but this fails to translate into innovation because HELIs lack confidence.)	Slack resources have an insignificant effect on both intrinsic and social motivation among LEHIs.	Slack resources enhance intrinsic motivation and lead to more innovation; the effect of tech slack is especially strong.
Recommendation	Offer HEHIs time and support slack, and motivate them in social terms.	Motivate HELIs in intrinsic terms; add slack, especially tech slack, and provide a "safe place to play."	Motivate LEHIs in intrinsic terms. There's no need to add slack, but do provide a "safe place to play."	Motivate in intrinsic terms; add slack, especially tech slack, and provide a "safe place to play."

A Decision Tree for Designing Slack Innovation Programs

This figure shows how employee motivation, the effects of slack, and recommended management actions change as employee expertise and self-assessed innovativeness vary.

they do it with an eye toward the impact the improvements will have on others or the organization and its mission.

In our sample, HEHI employees exhibited both intrinsic and social motivation. The motivations of the other three types of employees are more moderate. Non-HEHI employees have what appears to be an innovation confidence problem, perhaps because of their lack of expertise and/or an aversion to innovation. In terms of social motivation, they simply don't consider it feasible that they could help others by innovating. Intrinsic motivation remains a possibility, but, as we will explain shortly, managing in ways that activate intrinsic motivation for these three types of employees is a more complex matter than just offering them slack resources.

Time, Tech, and Support Slack

This brings us to the question of which kinds of slack resources work best with what types of employees. There are three kinds of slack resources that can be applied to employee innovation: *time*, that is, paid time intended for innovation that is carved out of employees' day-to-day schedules; *technology*, that is, access to hardware and software tools beyond those that employees need to do their regular jobs; and *support*, that is, access to the experts who can help employees pursue their ideas. The effectiveness of these three kinds of resources varies by the type of employee to whom they are given.

Time   Time may be the most common slack resource. Creativity research suggests that innovation does not often happen under pressure. Rather, producing something original requires doing something different from what you usually do, and that requires companies to subsidize time away from day-to-day work.

But does adding time slack actually enable most employees to produce innovations? Our results clearly show that adding this form of slack makes HEHI employees more innovative. Time slack programs send the message that innovation is important to the organization (in part by making it "safe" for employees to take time to explore) and thereby support and amplify the HEHI's social motivation. Amplifying social motivation also diminishes the HEHI's intrinsic motivation (by directing the inclination to innovate outward), but that doesn't matter much because HEHIs are essentially on autopilot when it comes to the intrinsic motivation to innovate and because a more outward focus can remind HEHIs that they shouldn't become isolated in their passion and should instead look for ways to connect their work to others'.

Adding slack time does not work that way for the other three types of employees. Because of their lack of expertise and/or their perceived lack of innovativeness, their social motivation to innovate does not increase when they are offered slack time. Our results show that adding slack time does amplify the intrinsic motivation to innovate in these employees to some degree (it is a confidence enhancer). But time to innovate in and of itself is not enough to produce innovation among non-HEHI employees.

Technology   Adding slack in the form of extra technology that is not strictly required to do a job provides additional opportunities for employees to explore innovative ideas and take risks at a relatively low cost. In our interviews, subjects mentioned databases, enterprise applications, and cameras that simulate 3-D as some of the technological resources that their companies had provided to them. We found that adding technological slack

does little for employees who see themselves as highly innovative (HEHIs and LEHIs). But access to new technologies does help
HELIs and LELIs become more comfortable with innovation. We
don't know why exactly; perhaps it's because they see the added
technology as a more reliable source of innovation than their
own abilities.

Support Support slack, the provision of technically expert
support personnel, such as engineers or programmers, to help
employees pursue innovation, came up often in our interviews.
Interviewees told us that easy access to support personnel bolstered their confidence in their ability to overcome obstacles
that arise during exploration and experimentation. According
to our analyses, support slack is very helpful to HEHIs. As with
time slack, this may be because management's decision to invest
in extra resources sends a signal that innovation is important to
the organization. That message bolsters their social motivation
to innovate and their sense that it is safe to spend time on innovation. Support slack also has a favorable impact on social motivation to innovate among HELIs, but because of their relatively
low confidence, it does not translate into additional innovation.
Support slack has no discernible effect on LEHIs, but it does bolster the intrinsic motivation of LELIs and is associated with a
modest increase in their tendency to innovate.

Making the Most of Slack Innovation Programs

Our findings suggest six issues for companies to consider in
designing and implementing slack innovation programs.

1. **Slack innovation programs are not one-size-fits-all undertak-
    ings.** Unless you have at least as many HEHI employees as, say,

Google, you're not going to get the same bang out of a slack innovation program as highly innovative companies do. To be effective, slack programs must be tailored to the expertise and inherent innovativeness of the people who actually work for a particular company. Moreover, the impacts of such programs will be different among different categories of employees, so calibrate your expectations appropriately.

2. **Encouraging employee innovation requires managerial support at all levels.** Senior leaders make the decision to offer slack innovation programs. But it's middle and front-line managers who must customize the allocation of slack and motivate employees to use it, based on their expertise and innovative inclinations. Democratize slack allocation decisions among these managers. and ensure that they are supportive and in sync with slack innovation programs at every level.

3. **Combine slack resources with appropriate motivational framing.** Motivational framing is crucial to the success of slack innovation programs. When allocating slack for HEHIs, social motivation will be the most effective means of ensuring that they keep their heads up and see the big innovation picture. For the three other employee types, the most effective messages to bundle with slack are those that appeal to intrinsic motivation and position innovation as a means of personal growth and fulfillment.

4. **Provide a "safe place to play" for employees who have low expertise and/or low self-assessed innovation.** Employees who don't have much expertise or don't regard themselves as very innovative need verbal assurances and other slack resources to feel safe in their explorations. If such support is not available, slack programs aimed at those populations of employees will not spawn innovation. Create a playful

environment that makes exploring and experimenting with ideas low risk, cheap, and fast.

5. **Employ the right kinds of slack for the right employees.** Time and support slack can be bundled to create a reinforcing effect, while tech slack only helps employees with low self-assessed innovation. For HEHIs, time and support slack bolster social motivation; for HELI and LELI employees, time and support slack bolster intrinsic motivation by boosting confidence and safety. However, none of the three kinds of slack is particularly effective with LEHI employees. This may be because such employees tend to use slack resources to build their on-the-job expertise instead of pursuing innovation.

   Managers also should note that people who start out as one type of employee can transform into another type as they gain expertise and elevate their perceptions of personal innovativeness. As a result, managing slack innovation programs is a dynamic and ongoing process.

6. **Design slack innovation programs for the type of innovation you want.** Adding slack can produce two types of innovation: internal innovations, which address work processes and deliver efficiency gains, and outcome innovations, which address process deliverables and directly affect customers. Offering slack resources to employees with low expertise and/ or low self-assessed innovation typically results in internal innovation.

   Offering slack resources to HEHIs, however, is more likely to produce outcome innovations, such as new products. Thus, you should keep your company's innovation needs in mind as you choose which employees to provide with slack resources and then seek to motivate those employees to deliver innovations.

Cut 'Em Some Slack

In 1959, Miles Davis, the legendary jazz trumpet player and band leader, brought six talented musicians into a New York City recording studio. Instead of asking them to play bebop, a complex style of jazz that pushed the limits of even the best players, he gave them much easier music. In doing so, Davis effectively freed up large swathes of their mental and musical capacity. You might say that he added slack, which, in turn, bolstered the group's ability to innovate.[5] The result was *Kind of Blue,* a masterpiece that pioneered a new kind of jazz called modal jazz. It remains, nearly 60 years later, the best-selling jazz album of all time.

Such is the potential payoff of adding slack resources to bolster innovation. But to realize that potential, leaders need to design, implement, and continuously manage slack innovation programs that match the needs of their workforces.

15

Need Motivation at Work? Try Giving Advice

Lauren Eskreis-Winkler and Ayelet Fishbach

At companies such as IBM and Motorola, mentorship and executive coaching are now standard parts of leadership development programs. People seeking wisdom from mentors is common not only in business, but in all facets of life.

But what if advice seekers are overlooking what they truly need? Psychologists have long known that people stumble on one particular class of problems, self-control problems, because they lack the motivation to transform knowledge into action.

Realizing this, we decided to turn the standard solution to self-control on its head: What if instead of seeking advice, we asked struggling people to give it? Across a series of experiments, we appointed populations struggling with self-control—everything from academic problems to money problems to health problems—to advise others on the very problems they were encountering. Although giving advice confers no new information to the advice giver, we thought it would increase the advice giver's confidence. Confidence in one's ability can galvanize motivation and achievement even more than actual ability.

In one study, we recruited a sample of unemployed individuals struggling to find a job. We asked these individuals to give job search advice to their equally deflated peers. Next, all participants read job search tips from The Muse, a professional career advice platform. After giving and receiving advice, 68% of unemployed individuals reported that giving advice made them feel more motivated to search for jobs than receiving advice.

This method proved a powerful motivator in the financial domain as well. Approximately 72% of people struggling to save money found giving advice more motivating than receiving tips from experts at America Saves. Likewise, 77% of adults struggling with anger management found giving anger management advice more motivating than receiving advice from professional psychologists at the American Psychological Association. Finally, 72% of adults struggling to lose weight found giving weight loss advice more motivating than receiving advice from a seasoned nutritionist at the Mayo Clinic.

Repeatedly failing to achieve one's goals saps confidence. For a number of reasons, giving advice may restore it. For example, simply being asked to provide advice implies to those advice givers that they possess, as opposed to lack, the ability in question. Giving advice prompts one to conduct a biased memory search by considering past successful behaviors in order to generate advice for others.

To test whether giving advice would affect behavior over time, we ran a three-week advice intervention in a Midwestern middle school. The school's 318 students were randomly assigned to be advice givers or advice receivers. Advice givers spent 38% more time on their homework than the advice receivers spent over the month following the intervention.

If giving advice motivates behavior among children and adults, in work and personal domains, then why is this activity so rare? When was the last time you told a demotivated colleague it would be a good idea for her to go and motivate others? When was the last time you appointed an employee who couldn't stop procrastinating to give time-management advice to someone else? Probably never. In our data we find that people overwhelmingly (erroneously) believe that both they and others will be more motivated by receiving advice than giving it. People falsely attribute failures in self-control to lack of knowledge, but lack of confidence—and by extension, motivation—are the true culprits.

There are clear, practical implications for management when it comes to motivation and advice. Employees struggling to maintain motivation at work ought to give advice as much or more than they receive it. For example, if an employee is experiencing a problem at work with time management, our research shows that this employee would benefit from being asked to counsel a colleague on how to help prioritize their tasks and manage their workload.

Our findings suggest a rich, undetected source of motivation lies in our midst. While managers should continue to assist employees with direct mentorship opportunities, they might also benefit from reframing the conversation, asking their reports to give others advice on the very problems they see their employees struggling with. Typically, past studies have not focused on this type of mentorship, but flipping the paradigm from advice seeking to advice giving offers much motivational promise.

16

Why People Believe in Their Leaders – or Not

Daniel Han Ming Chng, Tae-Yeol Kim, Brad Gilbreath, and Lynne Andersson

Leadership is the relationship between people who aspire to lead and those who choose whether or not to follow.[1] And it hinges on the leader's credibility, which is difficult to build and easy to lose. In recent years, numerous corporate executives—including the CEOs of BP, Wells Fargo, and Volkswagen—have learned that tough lesson through high-profile scandals that swiftly damaged their reputations.[2]

But what's at the heart of credibility? Two critical elements: perceived competence (people's faith in the leader's knowledge, skills, and ability to do the job) and trustworthiness (their belief in his or her values and dependability).[3] Such views are formed through direct and indirect observation of the leader's work and performance. And these perceptions are extremely important in a digital age, when vast amounts of information about people can be captured and scrutinized through technologies like smart sensors and artificial intelligence systems. Employees also seek assurance that those who are managing them and assessing their performance are competent and trustworthy.

Researchers have identified several broadly defined behaviors that influence whether leaders are perceived that way.[4] These behaviors include knowing oneself, appreciating one's constituents, affirming shared values, developing new capabilities, serving a purpose, and sustaining hope. However, not much has been written about concrete actions that enhance or harm a leader's credibility. Indeed, it's widely assumed that behaviors that don't increase credibility naturally decrease it. Research has begun to challenge this assumption,[5] but we had many unanswered questions, so we set out to learn more.

In several field studies, we explored the specific behaviors that affect how people assess their leaders' competence and trustworthiness and, in turn, their credibility. From this work, we have gleaned the following insights into what causes leaders to gain or lose credibility with their employees and what leaders who have lost credibility can do to regain it.

About the Research

We conducted several field studies over three years, using both quantitative and qualitative methods, to understand what affects leaders' credibility and how their credibility influences employee behaviors and organizational outcomes. The studies included both blue- and white-collar employees from different parts of the United States and with varying levels of formal education. We developed a comprehensive model of top managers' credibility through a field study involving 146 respondents: employees in a transportation company and evening MBA students employed in a variety of organizations. Analyzing the answers provided by our respondents, we identified leadership behaviors that generate perceptions of competence and incompetence, trustworthiness and untrustworthiness—factors that either underpin or undermine leader credibility.

To cross-validate and refine the set of behaviors, we conducted a second field study with 145 respondents: employees in a service-industry organization and a second set of evening MBA students. The respondents assessed the extent to which the behaviors identified in the first study indicated a leader's competence or incompetence and trustworthiness or untrustworthiness.

How Leaders Build Credibility

Based on input from employees we surveyed from a range of organizations, we found that leaders are perceived as competent when they place an emphasis on the future, on organizational outcomes, and on employees, as well as when they take action and launch initiatives, communicate effectively, and gain knowledge and experiences. At the same time, we identified several behaviors that point to trustworthiness. They include communicating and acting consistently, protecting the organization and employees, embodying the organization's vision and values, consulting with and listening to key stakeholders, communicating openly with others, valuing employees, and offering support to employees and stakeholders. Although scholars have already described many of these behaviors as signs of exemplary leadership[6] and credibility[7] in general, deeper analysis reveals specific actions that leaders can take to enhance their credibility.

Behaviors that project competence In the context of senior management, what are the best ways to emphasize the future and organizational outcomes and to take action and launch initiatives? One is to create clear plans for future success. This is different from simply stating a strategic vision or setting performance targets. It involves mapping out, in detail, how the organization will achieve its goals. Another way is to demonstrate sophisticated knowledge of industry trends and clear ideas about how the organization should respond to them. Still another approach involves actively predicting and preparing for upcoming changes by, say, making strategic investments in new technologies or markets. More than 80% of our respondents identified these behaviors as strong indicators of a leader's competence.

A sense of competence is enhanced when leaders work consistently to improve organizational structures and processes and maintain fiscally sound operations. These actions might include eliminating unnecessary reporting structures and spending, establishing new roles, or investing in technology that improves operational efficiency or business effectiveness.

It's often noted that true leaders are willing to take on big problems that others are reluctant to tackle. This sentiment was reinforced by more than 60% of our respondents, who told us that they saw leaders as competent when they were action-oriented and aggressive, when they took on issues or projects that needed to be addressed, and when they weren't afraid to make tough decisions.

Behaviors that project trustworthiness Consistent with previous research, we found that leaders are perceived as trustworthy when they communicate and behave in a consistent manner. To begin with, this means making decisions that aren't contradictory. But it also means behaving in a way that aligns with promises (both explicit and unspoken) that the company makes to employees and other stakeholders. By preemptively looking out for stakeholders' needs, executives can prevent stakeholder conflicts and organizational crises, as well as gain the trust of key stakeholder groups.

Another core behavior that can establish and enhance a leader's trustworthiness is to embody the organization's mission, both professionally and personally. Yvon Chouinard, the founder and former CEO of outdoor apparel company Patagonia, provides a good example. An avid outdoorsman and adventurer, Chouinard founded Patagonia with a specific mission: "Build the best product, cause no unnecessary harm, use business to inspire, and implement solutions to the environmental crisis."[8]

Throughout the company's 45-year history, Chouinard has lived and celebrated this mission. Employees are expected to use the company's products (some of which they can get for free) so that they are well informed about what they sell, and they are encouraged to participate in outdoor adventures to stay connected to the natural environment. And on a personal level, Chouinard makes time for mountain climbing, skiing, and other outdoor activities with friends and family.

How Leaders Erode Credibility

While prior research was less focused on factors that cause leaders to lose their credibility, employees in our field studies identified a number of red flags in both the competence and trustworthiness categories. Many leaders are unaware that they are acting in these ways or that such behaviors are damaging their credibility, so we will describe them here.

Behaviors that suggest incompetence   More than 80% of our respondents told us they view their top managers as incompetent when they display a lack of relevant job knowledge. Although people often assume that leaders are selected because of their knowledge, skills, and abilities, this isn't always the case. Leaders risk losing credibility quickly when they struggle to handle key tasks that are part of their job, have difficulty answering questions about the organization, or make decisions that don't align with the organization or its broader environment. An extreme example of this is Tony Hayward, BP's CEO during the 2010 Deepwater Horizon oil spill, who during the crisis repeatedly showed a lack of understanding of the accident's causes and severity and its devastating social and environmental consequences.

As we have noted, an important characteristic of competent leaders is that they take on big problems. So it makes sense that more than 70% of our respondents told us that they seriously question the competency of leaders who fail to take action or ignore problems. Commenting on the 2016 Wells Fargo scandal, in which bank employees opened 2 million accounts without customers' permission, Warren Buffett said that then-CEO John Stumpf's biggest mistake was his failure in the preceding years to address the underlying policies that triggered the scandal.[9] In essence, Stumpf showed the sort of laissez-faire approach to leadership that people often equate with incompetence.

One of the surest ways leaders raise questions about their competence, employees noted, is to create confusion among employees and other stakeholders. A particular example that more than 70% of our respondents cited is distributing incorrect information. Sometimes leaders do this without realizing it; sometimes they misrepresent the facts by trying to put a positive spin on difficult situations. Either way, people end up confused at best—and suspicious at worst. This is a high-stakes problem in today's business environment, where leaders are expected to handle information from numerous sources with great care and discretion.

Another behavior that undermines a sense of competence is giving contradictory information. The contradictions might come from different people on the leadership team or even from the same person. For example, Hayward was severely criticized for providing incorrect and inconsistent information during the Deepwater Horizon oil spill. He was quoted as stating that "the overall environmental impact of this will be very, very modest" in spite of clearly contradictory information.[10]

Finally, leaders can damage their reputations for competence when they ask for information and reports that don't seem

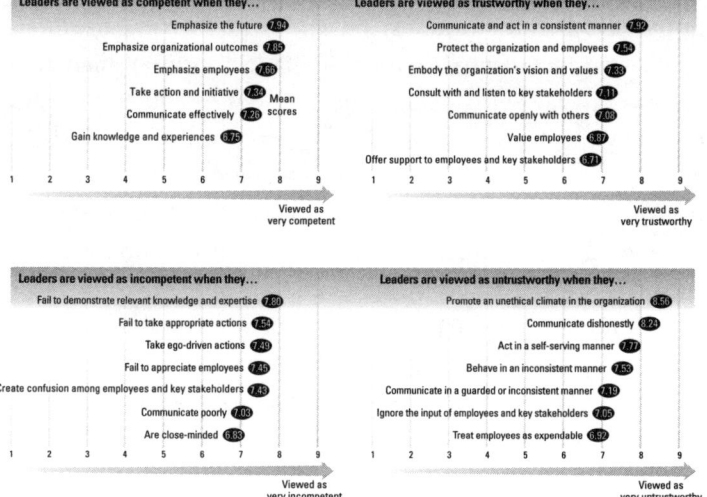

Actions that Build or Destroy Credibility

Using a nine-point scale, 145 employees in a range of organizations rated these leadership behaviors as indicators of competence and trustworthiness—traits that people associate with credibility.

These behaviors emerged as indicators of incompetence and untrustworthiness.

relevant or worthwhile. Such seemingly extraneous requests can cause confusion as to what the organization's priorities are, and employees may feel resentful about what they see as a waste of their time.

Behaviors that suggest untrustworthiness Since trust is so fundamental to the relationship between leaders and their constituents, behaviors that suggest untrustworthiness quickly undermine credibility. We identified several behaviors that one might think leaders would realize are detrimental and avoid doing. They include promoting an unethical climate within the organization by misappropriating resources (as Tyco International CEO Dennis Kozlowski did when he used company money to throw a birthday party for his wife in Sardinia); manipulating or even falsifying data to make things look better (as illustrated by the "creative bookkeeping" used by Enron and Arthur Andersen); and engaging in sexual harassment of, or illicit relationships with, employees. In fact, our respondents mentioned these behaviors frequently, with more than 80% indicating that the behaviors were very suggestive of an untrustworthy leader. Even if leaders don't act unethically themselves, they can suffer a serious loss of trust if they permit colleagues to act unethically. Leaders must uphold high ethical values to protect their organization and its people, or their followers and key stakeholders will lose faith in them.

Dishonest communication is another seemingly obvious way leaders hurt their trustworthiness. This goes beyond trying to paint something in the most favorable light possible. Leaders who relay false or inaccurate information or keep lots of secrets jeopardize their credibility, as do those who make promises without making any effort to fulfill them, for example, by saying "I'll get back to you," but never doing so.

Our research also revealed that self-serving behaviors can undermine employees' trust in their leaders. These include bending the rules to privilege themselves or close associates, making decisions based on their own self-interest rather than what's best for the organization, urging employees to make material sacrifices while wasting the organization's resources on perks for themselves, and taking credit for the achievements of others.

Leaders who openly ignore the opinions of employees and key stakeholders are also perceived as untrustworthy. Specific examples given by our respondents include making unilateral decisions and casually rejecting others' requests without due consideration. More than 60% of our respondents identified these behaviors as strong indicators of a leader's untrustworthiness.

Although it's clear that leaders lose credibility when they display incompetence or untrustworthiness, scholars have found that employees are much more tolerant and forgiving of an incompetent leader than they are of an untrustworthy leader.[11] They believe that incompetent leaders can at least *try* to become more competent, whereas untrustworthy leaders can't easily become more trustworthy.

Insights for Leaders

So far, we've described behaviors that build or erode credibility so that leaders can more accurately assess how they're comporting themselves and how others see them. Next, we will share a few insights from our analysis so that leaders will better understand how to avoid losing credibility or, if they've already lost it, how to get it back.

1. **The behaviors that help you gain or lose credibility aren't always mirror images of each other.** Some behavioral indicators for competence versus incompetence or trustworthiness versus untrustworthiness are mirror images; for example, taking action can suggest competence, while failure to do so can suggest incompetence. More often, though, behaviors that cause employees to perceive senior managers as competent or trustworthy aren't inversely related to those that convey incompetence or untrustworthiness. While emphasizing the future can indicate competence, for instance, it's not as though emphasizing the past is a sign of incompetence.

   Because many of the perceived behaviors of competence and trustworthiness are asymmetric, avoiding behaviors that make leaders *lose* credibility doesn't automatically help them *gain* credibility. Indeed, even when they engage in behaviors that enhance credibility, leaders might still lose credibility by engaging in behaviors that indicate incompetence and untrustworthiness. So it's important to consider the full range of indicators when trying to gauge how others see you as a leader.

2. **Sometimes positive information carries more weight than negative information – and vice versa.** Scholars who study trust have found that people tend to weigh positive information more heavily than negative information with regard to competence. However, people weigh negative information more heavily than positive information when it comes to trustworthiness.[12] A single competent act may be seen as a reliable signal of competence, but a single incompetent act is more likely to be dismissed as an outlier. On the other hand, people tend to attach more significance to a single untrustworthy act than to a single trustworthy act. This suggests that

leaders can gain credibility by performing one action that projects competence, such as creating a clear vision for the organization's future. But they can easily lose credibility by engaging in an untrustworthy action, such as manipulating data to mislead others.

3. **Overcoming the loss of credibility is difficult – but possible.** Any behavior that causes employees to attribute incompetence and untrustworthiness to top management, either alone or in combination with other behaviors, can have negative repercussions that are tough to recover from. As noted earlier, employees are less tolerant of untrustworthy behaviors than of incompetent behaviors. Partly for that reason, it's more difficult to regain credibility once it's lost than to build credibility in the first place.[13] But it can be done.

In the wake of the financial crisis in the late 2000s, for example, some of the questionable practices of major U.S. financial services companies were exposed in the media and scrutinized by the public. Many companies responded by making relatively modest changes to executive compensation or governance practices. However, James Gorman, CEO of Morgan Stanley, took the opportunity to thoroughly review the company's practices regarding compensation, compliance, and risk management. He refocused the company's culture on sustainable long-term performance goals, ethical management of resources, and a renewed emphasis on the interests of clients, and earned praise and trust from employees and other stakeholders. But many leaders don't have that experience when trying to regain credibility after their (or their organization's) trustworthiness has been questioned. Such turnarounds often require companies to install new leadership, as in the wake of the Wells Fargo customer account scandal, which forced CEO Stumpf to resign.

To regain lost credibility, leaders must reestablish positive expectations, which means they must repeatedly engage in trustworthy acts, since a single act won't mean much. They also need to overcome negative expectations that stem from their incompetent and untrustworthy behaviors by emphasizing the specific behaviors that project competence and trustworthiness.

17

Building an Ethically Strong Organization

Catherine Bailey and Amanda Shantz

When German car manufacturer Volkswagen was caught cheating on its diesel emissions testing regime a few years ago, the subsequent scandal launched numerous lawsuits, cost billions of dollars in fines, and severely harmed the company's reputation. The actions—and inaction—of dozens of employees at all levels, across divisions and countries, contributed to this disaster, including the software engineers who designed the cheating device, the workers who installed it, the managers who approved the fitting and testing, and the members of the senior leadership team who either orchestrated the scam or simply turned a blind eye.[1]

Of course, VW isn't an isolated example. Consider the costly lapses in judgment at Wells Fargo,[2] for instance, and at Samsung Electronics.[3] Why do such scandals continue, despite the clear moral and financial imperatives for ethical action? And—perhaps more important—what can be done to change matters?

Although some argue that people are innately inclined to behave unethically out of self-interest,[4] our research reveals that organizational ethics matter significantly to most employees and managers, and that people want to work for employers whose values and principles are aligned with their own. This suggests that ethical employers are likely to attract and retain ethical employees.[5] What's more, research has shown a link between ethical leadership and task performance, organizational citizenship, and other productive work behaviors[6]—companies have many compelling reasons to address ethical failings at the earliest opportunity. The urgency is all the greater in this digital age, since businesses must continually make rapid, high-stakes choices about how to handle sensitive customer and employee data.

To uncover the reasons behind persistent unethical conduct, we asked employees at five U.K. organizations—a national government department, a nationwide retailer, a nonprofit in the social services sector, a county-level police force, and a construction company—to tell us about their experiences of both ethical and unethical practices on the part of their colleagues, line managers, and senior executives.[7] We found that the ethical tone of an organization is the cumulative outcome of how its members address daily ethical dilemmas as they go about their work. Over time, a consistent mishandling of these micro-level issues can spiral into macro-level corporate scandal. Here, we discuss several murky areas that employees must navigate and ways that organizations can help them make ethical choices day to day.

About the Research

To inform our study design, we carried out a detailed analysis of research over the past 25 years on ethical leadership and decision making. We then surveyed a representative sample of 1,319 workers in the United Kingdom and conducted in-depth case studies in five U.K. organizations: a central government department with 18,000 employees, a nationwide retailer with 31,000 employees, a nonprofit in the social services sector with 1,100 permanent and 300 temporary staffers, a police force of 3,000 officers plus civilian staffers, and a construction company with 6,900 employees.

In four of the organizations, we surveyed 1,033 employees and their 524 line managers. Across all five, we conducted 46 face-to-face interviews, held 16 focus groups with a total of 79 participants, and analyzed company documentation such as human resources policies and statements of mission, vision, and values.

Daily Dilemmas That Trip People Up

When employees don't have a shared understanding of events that unfold around them, what constitutes an ethical response, and the consequences of behaving otherwise, it often means the organization has created an ethically weak situation for them. People essentially become free agents, behaving idiosyncratically in the absence of clear, strong norms. (An ethically strong situation, in contrast, is one in which "the right thing to do" is clearly communicated to employees and people have the motivation and ability to behave in ways that are consistent with the organization's ethical code.[8]) In the case of VW, an ethically weak situation was allowed to develop over many years, as senior executives prioritized market share over environmental and legal concerns in one judgment call after another.

Here are the daily dilemmas we found that tend to muddy the ethical waters for individuals in decisions both large and small.

Ethical disconnect Sometimes employees observe a gap between their personal ethics and those of the wider organization, and that makes them uneasy. An abundance of studies show that people want to fit in at work[9]—but it's not just a fit with the requirements of the job or even a fit with the organization's culture that matters. New research is beginning to show that people have a strong desire to gain a sense of moral fit as well.[10]

Because they feel this deep-seated need, they're desperate to close the gap between their own ethics and those of their organization. When they struggle to do so, they often withdraw and may quit their jobs altogether. One manager told us, "I've worked in businesses that I didn't stay in very long because of the ethics and the culture. I didn't feel comfortable." This sentiment is echoed by many.

Conflicting stakeholder needs   Every organization has a range of stakeholders affected by its decisions, including employees, suppliers, clients, senior managers, the local community, wider society, and even the environment.[11] Organizations may have an explicit approach to balancing these competing needs—but that may not be the same as the implicit approach that employees witness every day.

When we asked employees and their leaders to rank the order in which stakeholders "matter" in important decisions, consensus was rare. As one employee in the retail sector said, "Even though we've got a vision and we've got an ethical policy framework, I personally feel very strongly that [in practice] it's shareholder, company, colleague, in that order."

When groups of stakeholders lobby for special treatment, the situation becomes even more complex. For the nonprofit we studied, a core challenge was figuring out how to handle large donations that are linked to requests for preferential care of the donors' relatives. One manager told us, "Sometimes, the choices we have to make are not overtly compromising, but they can make things difficult—people asking for access to services when

When employees choose to stay quiet – even with good intentions – alternate viewpoints are silenced, levels of engagement and commitment are likely to diminish, and others note that failure to challenge is the norm.

they're not entitled to them, or people jumping the queue." Managers must weigh the monetary worth of the donation against the nonprofit's values of integrity, fairness, and transparency.

While the nonprofit solved this dilemma by refusing to provide preferential treatment in exchange for donations, situations vary, and what is right for one organization may not be right for another. Even different departments within the same organization face competing priorities when having to choose between stakeholder groups. However, each time an employee or a leader makes a decision that implicitly or explicitly favors one stakeholder group over another, it sends a message to other employees about what really matters—and whose interests the organization is willing to sacrifice.

Not knowing whether (or how) to speak up   Witnessing unethical conduct by a colleague or superior forces people to decide: Do I take this further? If so, how? And what will be the consequences for me and for others?

Often, whether or not people challenge unethical behavior depends on the nature of the infraction, the setting within which it takes place, the seniority and roles of those involved, and the potential risks of challenging the behavior. Some ethical breaches are especially difficult to challenge; in many cases, staff may be unwilling to challenge upward. One government manager seemed to have realized this, saying, "I'm quite an outspoken person, and nobody has ever challenged my behavior, even though in some circumstances I recognize that I perhaps go a bit too far."

Possible responses include staying silent, taking the individual aside and discussing the matter privately, calling the person out in front of others, reporting the matter to senior staff, or reporting it anonymously via a whistle-blowing or antiharassment program.

Some employees we spoke with described instances when they chose to stay silent. Discussing an event when bonuses were awarded to everyone except the hourly workers on the front line, one retail employee said:

It did feel desperately uncomfortable, but in the end you either rise up as a whole population and say, "No, this isn't right, none of us are taking bonuses," or you become an outlier and a single person saying, "I don't want my bonus, I'm going to give it to charity," or you say nothing. I didn't say anything.

And a junior police officer told us:

If you and I were constables and I'd seen you behave in an unethical way and challenged you about it, that could cause bad feeling. But then if you and I went out and faced somebody going crazy with a knife, I'd need to know you'd have my back. It's not like working in an office. You might be relying on that person to save your life.

When employees choose to stay quiet—even with good intentions—alternate viewpoints are silenced, levels of engagement and commitment are likely to diminish,[12] and others note that failure to challenge is the norm.

Conversely, in the construction company, an employee was comfortable publicly challenging a colleague for the use of sexist language; when the perpetrator apologized immediately, the interaction sent a positive message to others about how to handle such situations.

Ethics versus expediency Another challenge is deciding what to do when the ethical solution to a problem is not the expedient solution—often because there aren't enough hours, dollars, or people to make the ethical solution happen. As one retail manager put it:

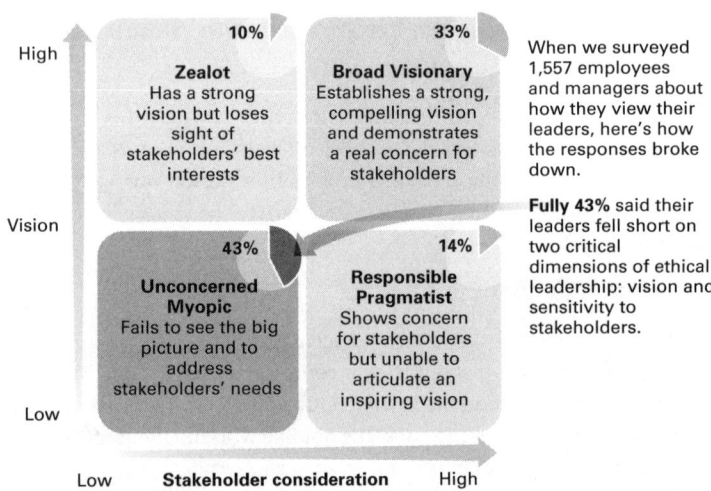

When we surveyed 1,557 employees and managers about how they view their leaders, here's how the responses broke down.

Fully 43% said their leaders fell short on two critical dimensions of ethical leadership: vision and sensitivity to stakeholders.

Two Critical Elements of Ethical Leadership

The number of CEOs sacked for ethical misconduct has risen 36% in the last five years,[13] including such high-profile examples as Yahoo CEO Scott Thompson, United Airlines CEO Jeff Smisek, and LendingClub founder Renaud Laplanche. But the problem of unethical behavior can't be "solved" simply by firing senior leaders who behave badly.

To bring about lasting change, organizations must invest in "distributed" ethical leadership. That is, they must hire and cultivate leaders at all levels who promote ethical behavior. Two essential ingredients are a strong vision and a deep commitment to stakeholders.

Our research shows that employees who see their managers as ethical leaders are more satisfied with their work, are more willing to go the extra mile, find the work that they carry out has significance in the broader scheme of things, and are less likely to quit. Unfortunately, most organizations aren't poised to reap those rewards.

I think our ethics as a business are very, very good. Where we get the frustration is when we want to do the right thing with our people, but actually the resource levels that we're asked to work on make it impossible sometimes.

In the context of the police, this kind of problem meant that officers had to make choices about which crimes to investigate, causing "a huge amount of strain and stress to officers because they can't do the job they're trained to do, that they're paid to do, that they want to do, and is the reason why they joined in the first place," according to a leading officer on the force.

Call to Action

No organization is free of these dilemmas, but they can be managed. Our research and analysis suggest that the following six steps can help leaders set an ethically strong tone so that employees are better equipped to make the right choices day to day.

1. **Acknowledge ethical ambiguity.** Many organizations do not recognize or discuss ethically tricky situations their managers and employees face. This drives individuals to internalize their decision-making processes—which can create a slippery slope.

   In the police force we studied, even though leading officers were well aware that budget cuts meant increased workloads and longer hours for the rank and file, they had not openly acknowledged these pressures with their staff and how they might affect day-to-day decision making—preventing an authentic dialogue about the problems or possible solutions. One leading officer said, "We are really struggling, and we're not admitting that to people on the ground." Officers and staff felt the pressure but, given the lack of open discussion, assumed that senior leaders did not care.

In organizations with a culture of transparency, people are more inclined to seek to understand the underlying rationale for decisions. This has a positive effect on ethical decision making because values are exposed when they are openly discussed rather than inferred from town hall meetings or company documentation. At the nonprofit we studied, one executive noted: "You know, I've worked for places where things are done behind closed doors and you don't really understand the reasons. I think here, whatever initiatives are being run, it's done very openly. We don't make decisions in hiding; we make decisions in a very consultative way." So when its employees wonder how, for instance, to respond to a donor who requests a service, it's easier for them to make that call, because they have a clear understanding of the organization's ethical values and are confident they can go to their managers for clarification or support without fear of being negatively judged.

2. **Clarify the ethical trade-offs.** Another important step is to explicitly clarify how employees should balance the needs of different stakeholder groups.

   Most decisions will affect more than one set of stakeholders. Although the needs of all groups can sometimes be met, trade-offs are usually necessary. When employees are not sure how to manage this tension, unethical approaches can develop.

   In the retail company, leaders paid lip service to meeting customers' needs above all others, but their behavior wasn't always consistent with that message, which created confusion. Employees reported that decision making was more often governed by immediate profit considerations and key performance indicators. Some felt a degree of cynicism toward the company's "customer first" rhetoric, believing

Another challenge
is deciding what
to do when the
ethical solution to a
problem is not the
expedient solution –
often because there
aren't enough hours,
dollars, or people
to make the ethical
solution happen.

that in practice senior executives were more concerned about hitting performance and sales targets by persuading customers to buy add-on products and services than about caring for the customer or providing excellent customer service. One employee said: "You always have that tagline at the end, 'The customer comes first,' but at the end of the day it's a business and the people at the top know we need to hit a KPI figure."

Confusion about whose needs to prioritize can be compounded when an organization has been through a series of mergers or takeovers that bring together different ethical climates. In these cases, leaders have an especially significant role in establishing a consistent ethical framework and guidelines for balancing stakeholder interests.

Providing employees with a clear statement of vision can help them weigh competing concerns and make appropriate trade-offs. In the police force, for instance, a widely shared "Plan on a Page" helped officers understand policing priorities (such as child abuse and exploitation, modern slavery, and violence) and provided guidance on serving the needs of the community (by putting the victim first and communicating effectively with the public) while also making the most efficient use of resources.

3. **Ensure role modeling from the C-suite down.** Employees observe how leaders actually handle ethical dilemmas, rather than what they say about ethics, and will infer the organization's real priorities accordingly. VW is a case in point: Though senior executives claimed to care about "clean diesel," they apparently both condoned deliberate cheating on emissions tests and encouraged employees to hide or destroy its evidence.[14]

When the senior team sends mixed ethical signals, mid-level managers may pick and choose what to follow. These mixed signals cascade through each level of the organization. As one employee in the construction business said, "If your direct line manager isn't setting an example for you, it detracts from the message that the business is giving."

We did find that the ethical conduct of midlevel managers can compensate for mixed messages from the C-suite, slowing or even reversing the development of an ethically weak situation. In the retail business, for example, the staff talked positively of the "family atmosphere" and shared values within individual stores and regions that counteracted the dominant "cost control" messages from the head office. However, a much more reliable approach is to set the desired example at the top. The nonprofit fostered an ethically strong situation by clearly showing how core ethical dilemmas should be resolved: When a company bidding to work with it asked one of the nonprofit's trustees to put in a good word for it, its leaders immediately ruled out the company as a partner due to a misalignment of ethical values.

4. **Embed ethics in corporate policies and programs.** Ethically strong situations are developed in settings with robust codes of conduct and policies for enforcing those codes.[15] Such policies should include clear rules about bullying, harassment, and whistle blowing. And they should be conveyed and reinforced through onboarding, leadership development, and other training programs.

Without formalized policies around ethics, efforts to create an ethically strong situation will most likely founder. As one police officer said, they "help people understand why we

Ethically strong situations are characterized by the presence of a transcendent cause that unites the organization behind a vision and set of values that go beyond self-interest.

need to behave, act, do things in a certain way, and what the consequences are for *not* doing that."

Although corporate policies and programs alone will not eliminate unethical practices,[16] their existence is essential. For example, at the nonprofit we examined, employees were frequently confronted with ethical dilemmas when working with clients, such as how to assess mental capacity or how to manage end-of-life issues and determine appropriate levels of treatment and support. The organization helps its employees make ethical decisions by developing clear policies on approaches to care and providing training that specifically focuses on such challenges.

Similarly, at the construction company, part of the recruitment process involves matching applicants' ethical values with those of the business. It has also adopted a code of conduct and a formal framework called "What Good Looks Like" to guide employee behavior. Training on topics such as how to deal with anticompetition risks and health and safety issues is compulsory for line managers, and an online system allows for logging any health and safety issues as they arise. Although employees sometimes feel that these processes slow decision making, they provide clarity and "consistency, and people know what is expected of them," according to a frontline manager.

5. **Empower individuals to handle ethical breaches.** Ethical breaches will inevitably arise, of course—whether through error, neglect, or deliberate action. But ethically strong organizations explicitly say how people should deal with them when they do occur, in addition to trying to prevent them in the first place. Employees at all levels then feel more

empowered—and obliged—to call out bad behavior, even when doing so may be difficult. For example, employees in the construction company are required to challenge decisions and actions that could compromise the health and safety of employees and customers alike. One manager said that the culture around this is so strong that "in extreme circumstances, people have lost their jobs because they haven't followed through on what really is their duty to either challenge it there and then or report it later to make sure remedial action can be put in place."

In ethically weak organizations, challenging people's behavior is not the norm. Sometimes employees fear retribution, because they do not see others around them raising questions. Or they may feel that no action will be taken if they do speak up.[17] Sadly, that assumption isn't necessarily unfounded. While some VW employees apparently did challenge the use of "defeat devices" designed to cheat the emissions tests, their concerns were ignored.[18] So far, the evidence suggests that more than 40 VW employees in different roles and at varying levels of seniority were implicated in the diesel emissions scandal.[19] Had individuals felt empowered to challenge ethical breaches, perhaps the scandal could have been contained before erupting on such a massive scale.

6. **Embrace a higher cause.** Finally, ethically strong situations are characterized by the presence of a transcendent cause that unites the organization behind a vision and set of values that go beyond self-interest. One employee called this "the vision that brings you back tomorrow."

The nonprofit's transcendent cause is to provide care and support for the community; for the police, it is to keep the community safe from harm. The construction company's ethical vision of sustainability translates into protecting the environment as well as safeguarding its employees and customers. As one of its managers told us, "A lot of practices in our industry do create harm for the planet, and so we're trying to reduce our CO_2 emissions."

When a company's mission or vision is unclear or divorced from ethics, or, as a senior leader at the retail organization said, when "the 'why' is missing" altogether, an opportunity to provide guidance is lost and an ethically weak situation develops. But an overarching sense of purpose creates a context within which micro-level ethical dilemmas can be resolved.

Conclusion

Setting the stage for ethical behavior isn't just a top-down exercise—though clear direction and positive role modeling from senior executives do help. Organizations must also consider the daily ethical dilemmas that their managers and employees face and give them the tools to make good choices. This involves regularly checking in to ensure that codes of conduct are clearly articulated and upheld—and imposing consequences when they are not.

No company is immune from ethically questionable decision making. But by openly acknowledging and carefully managing murky situations that come up again and again, organizations become much less susceptible to egregious lapses in judgment—and less likely to incur the associated reputational and financial costs.

Acknowledgments

The authors would like to thank the Chartered Institute of Personnel and Development for the funding that supported this research and the Involvement and Participation Association for its assistance in gathering data.

Contributors

Lynne Andersson is an associate professor of human resource management at Temple University's Fox School of Business in Philadelphia, Pennsylvania.

Robert D. Austin is a professor of information systems at Western University's Ivey Business School in London, Ontario, Canada, and an affiliated faculty member at Harvard Medical School in Boston, Massachusetts.

Catherine Bailey is a professor of work and employment at King's Business School at King's College London.

Kathryn M. Bartol is the Robert H. Smith Professor of Leadership and Innovation at the Robert H. Smith School of Business at the University of Maryland, College Park.

Daniel Han Ming Chng is an associate professor of management at China Europe International Business School in Shanghai, China.

Chris DeBrusk is a partner in the financial services and digital practices of Oliver Wyman, a global management consulting firm in New York City.

Arati Deo is managing director, machine learning and AI, for Accenture Technology Services in Bengaluru, India.

Kishore Durg is senior managing director, growth and strategy lead, and global testing services lead for Accenture Technology Services in Bengaluru, India.

Lauren Eskreis-Winkler is a postdoctoral researcher at the Wharton School of the University of Pennsylvania in Philadelphia, Pennsylvania, studying motivation and achievement.

Mallika Fernandes is associate director, AI testing, for Accenture Technology Services in Bengaluru, India.

Ayelet Fishbach is the Jeffrey Breakenridge Keller Professor of Behavioral Science and Marketing at the University of Chicago's Booth School of Business in Chicago, Illinois, studying motivation and decision making.

Fritz Fleischmann is the William R. Dill Governance Professor in the Arts and Humanities Division at Babson College in Wellesley, Massachusetts.

Kristen Getchell is a visiting associate professor and director of rhetoric in the Arts and Humanities Division at Babson College in Wellesley, Massachusetts.

Bhaskar Ghosh is group chief executive of Accenture Technology Services.

Brad Gilbreath is a professor of management at Colorado State University–Pueblo Hasan School of Business in Pueblo, Colorado.

Rob Gleasure is a lecturer in business information systems at Cork University Business School in Cork, Ireland.

Sergey Gorbatov is adjunct professor at IE Business School and director and general manager of development at AbbVie in Madrid, Spain.

Lynda Gratton is a professor of management practice at London Business School and director of its Human Resource Strategy in Transforming Companies program. She is coauthor of *The 100-Year Life: Living and Working in an Age of Longevity* (Bloomsbury, 2016).

N. Sharon Hill is an associate professor of management at the George Washington University School of Business in Washington, D.C.

Beth Humberd is an assistant professor of management at the University of Massachusetts Lowell's Manning School of Business.

Bala Iyer is the dean of faculty and a professor in the Technology, Operations, and Information Management Division at Babson College in Wellesley, Massachusetts.

Tae-Yeol Kim is the Philips Chair in Management at China Europe International Business School in Shanghai, China.

Frieda Klotz is a freelance journalist and correspondent for *MIT Sloan Management Review*.

Angela Lane is vice president, talent, at AbbVie in Chicago, Illinois.

Scott Latham is an associate professor of strategy at the University of Massachusetts Lowell's Manning School of Business.

Thomas W. Malone is the Patrick J. McGovern Professor of Management, a professor of information technology, and a professor of work and organizational studies at the MIT Sloan School of Management in Cambridge, Massachusetts, as well as the founding director of the MIT Center for Collective Intelligence. He is the author of *Superminds: The Surprising Power of People and Computers Thinking Together* (Little Brown, 2018), from which this article is adapted.

Daniel McDuff is an AI researcher at Microsoft in Redmond, Washington.

Paul Michelman is editor in chief of *MIT Sloan Management Review*.

Carsten Lund Pedersen is a postdoctoral researcher in the Department of Strategy at Copenhagen Business School in Frederiksberg, Denmark, where he researches strategy, employee autonomy, and business development in a digital age.

Alain Pinsonneault is the Imasco Chair of Information Systems and James McGill Professor in the Desautels Faculty of Management at McGill University in Montreal, Canada.

Yasser Rahrovani is an assistant professor of information systems at Western University's Ivey Business School in London, Ontario, Canada.

Fabrizio Salvador is a professor of operations management at IE Business School in Madrid, Spain.

Amanda Shantz is an associate professor in human resources and organizational behavior at Trinity Business School at Trinity College Dublin, Ireland.

Antti Tenhiälä is an associate professor of operations management at IE Business School in Madrid, Spain.

Jan vom Brocke is the Hilti Endowed Chair of Business Process Management and director of the University of Liechtenstein's Institute of Information Systems in Vaduz, Liechtenstein.

Eoin Whelan is a lecturer in business information systems at National University of Ireland Galway.

Notes

Chapter 2

1. N. G. Carr, *The Shallows: What the Internet Is Doing to Our Brains* (New York: W. W. Norton & Co., 2010).

2. S. E. Page, *The Difference: How the Power of Diversity Creates Better Groups, Firms, Schools, and Societies* (Princeton: Princeton University Press, 2007).

3. R. H. Thaler, *Misbehaving: The Making of Behavioral Economics* (New York: W. W. Norton & Co., 2016).

4. P. E. Tetlock and D. Gardner, *Superforecasting: The Art and Science of Prediction* (London: Penguin Random House, 2015).

5. A. Pentland, *Social Physics: How Social Networks Can Make Us Smarter* (New York: Penguin, 2014).

6. E. Pariser, *The Filter Bubble: What the Internet Is Hiding from You* (New York: Penguin, 2011); and E. Zuckerman, *Digital Cosmopolitans: Why We Think the Internet Connects Us, Why It Doesn't, and How to Rewire It* (New York: W. W. Norton & Co., 2013).

7. J. Diamond, *Guns, Germs, and Steel: The Fates of Human Societies* (New York: W. W. Norton & Co., 1997).

8. P. A. Gloor and S. M. Cooper, "The New Principles of a Swarm Business," *MIT Sloan Management Review* 48, no. 3 (spring 2007): 81–84; and S. Parise, E. Whelan, and S. Todd, "How Twitter Users Can Generate Better Ideas," *MIT Sloan Management Review* 56, no. 4 (summer 2015): 21–25.

9. C. Newport, *Deep Work: Rules for Focused Success in a Distracted World* (New York: Grand Central Publishing, 2016).

10. S. Pillay, "Your Brain Can Only Take So Much Focus," *Harvard Business Review*, May 12, 2017, https://hbr.org/2017/05/your-brain-can-only-take-so-much-focus.

11. J. Coleman, "The Best Strategic Leaders Balance Agility and Consistency," *Harvard Business Review*, January 4, 2017, https://hbr.org/2017/01/the-best-strategic-leaders-balance-agility-and-consistency. Although the two arguments are related, focus in attention is not necessarily the same as consistency, and agility is not necessarily the same as distraction. For instance, being consistent in action can be a rather mindless endeavor that does not necessarily need focus in attention. And being agile can necessitate very high focus of attention, making the interplay intricate. Moreover, this aspect is very closely related to "ambidexterity." See C. A. O'Reilly III and M. L. Tushman, "Organizational Ambidexterity in Action: How Managers Explore and Exploit," *California Management Review* 53, no. 4 (August 2011): 5–22.

12. M. T. Hansen, "Ideo CEO Tim Brown: T-Shaped Stars: The Backbone of Ideo's Collaborative Culture," *Chief Executive*, January 21, 2010, https://chiefexecutive.net/ideo-ceo-tim-brown-t-shaped-stars-the-backbone-of-ideoaes-collaborative-culture__trashed/.

13. R. A. Guth, "In Secret Hideaway, Bill Gates Ponders Microsoft's Future," *Wall Street Journal*, March 28, 2005, https://www.wsj.com/articles/SB111196625830690477.

14. P. Galanes, "The Mind Meld of Bill Gates and Steven Pinker," *New York Times*, January 27, 2018, https://www.nytimes.com/2018/01/27/business/mind-meld-bill-gates-steven-pinker.html; and B. Gates, *GatesNotes* (blog), http://www.gatesnotes.com.

15. "The Buffett Formula: Going to Bed Smarter Than You Woke Up," *Farnam Street* (blog), May 15, 2013, https://fs.blog/2013/05/the-buffett-formula-how-to-get-smarter/.

16. Many of the practical solutions for overcoming the distraction–focus paradox (dealing with attention) relate to better balancing agility and consistency. See Coleman, "The Best Strategic Leaders."

Chapter 6

1. R. Brooks, "Artificial Intelligence Is a Tool, Not a Threat," *Rethink Robotics* (blog), November 10, 2014, https://www.rethinkrobotics.com/blog/artificial-intelligence-tool-threat/.

2. David Ferrucci, email to the author, August 24, 2016. Ferrucci led the IBM team that developed Watson.

3. S. Armstrong and K. Sotala, "How We're Predicting AI—or Failing To," in *Beyond AI: Artificial Dreams*, ed. J. Romportl, P. Ircing, E. Zackova, M. Polak, and R. Schuster (Pilsen, Czech Republic: University of West Bohemia, 2012): 52–75.

4. N. Bostrom, *Superintelligence: Paths, Dangers, Strategies* (Oxford, United Kingdom: Oxford University Press, 2014).

5. M. Minsky, *Society of Mind* (New York: Simon and Schuster, 1988).

6. L. Biewald, "Why Human-in-the-Loop Computing Is the Future of Machine Learning," *Data Science* (blog), November 13, 2015, https://www.computerworld.com/article/3004013/robotics/why-human-in-the-loop-computing-is-the-future-of-machine-learning.html.

7. H. J. Wilson, P. Daugherty, and P. Shukla, "How One Clothing Company Blends AI and Human Expertise," *Harvard Business Review*, November 21, 2016, https://hbr.org/2016/11/how-one-clothing-company-blends-ai-and-human-expertise.

8. S. Wininger, "The Secret Behind Lemonade's Instant Insurance," November 23, 2016, https://stories.lemonade.com/the-secret-behind-lemonades-instant-insurance-3129537d661.

9. A. Kittur, B. Smus, S. Khamkar, and R. E. Kraut, "CrowdForge: Crowdsourcing Complex Work," in *Proceedings of the 24th Annual ACM Symposium Adjunct on User Interface Software and Technology*, ed. J. Piece, M. Agrawala, and S. Klemmer (New York: ACM Press, 2011); and "CrowdForge: Crowdsourcing Complex Tasks," *Boris Smus* (blog), February 2, 2011, https://smus.com/crowdforge/.

10. T. W. Malone, J. V. Nickerson, R. Laubacher, L. H. Fisher, P. de Boer, Y. Han, and W. B. Towne, "Putting the Pieces Back Together Again: Contest Webs for Large-Scale Problem Solving," in *Proceedings of the ACM Conference on Computer-Supported Cooperative Work and Social Computing*, Portland, Oregon, February 25–March 1, 2017.

11. J. Wolfers and E. Zitzewitz, "Prediction Markets," *Journal of Economic Perspectives* 18, no. 2 (2004): 107–126.

12. J. Wolfers and E. Zitzewitz, "Interpreting Prediction Market Prices as Probabilities," working paper W12200, National Bureau of Economic Research, Cambridge, Massachusetts, May 2006.

13. V. Granville, "21 Data Science Systems Used by Amazon to Operate Its Business," *Data Science Central* (blog), November 19, 2015, https://www.datasciencecentral.com/profiles/blogs/20-data-science-systems -used-by-amazon-to-operate-its-business.

14. P&G sold the Pringles business to Kellogg in 2012. For a description of the invention of the process for printing on Pringles, see L. Huston and N. Sakkab, "Connect and Develop: Inside Procter & Gamble's New Model for Innovation," *Harvard Business Review* 84 (March 2006): 58–66.

15. Martin Reeves and Daichi Ueda use the term "integrated strategy machine" to describe a somewhat similar idea. But unlike their article, this article focuses more on how large numbers of people throughout the organization and beyond can be involved in the process and on the specific roles people and machines will play. See M. Reeves and D. Ueda, "Designing the Machines That Will Design Strategy," *Harvard Business Review*, April 18, 2016, https://hbr.org/2016/04/welcoming-the-chief -strategy-robot.

Chapter 9

1. "Joint Commission Center for Transforming Healthcare Releases Targeted Solutions Tool for Hand-Off Communications," *Joint Commission Perspectives* 32, no 8 (August 2012): 1–3.

2. A. Smaggus and A.S. Weinerman, "Handover: The Fragile Lines of Communication," *Canadian Journal of General Internal Medicine* 10, no. 4 (December 2015): 15–19.

3. M. B. Lane-Fall, R. S. Beidas, J. L. Pascual, M. L. Collard, H. G. Peifer, T. J. Chavez, M. E. Barry, J. T. Gutsche, S. D. Halpern, L. A. Fisher, and F. K. Barg, "Handoffs and Transitions in Critical Care (HATRICC): Protocol for a Mixed-Methods Study of Operating Room to Intensive Care Unit Handoffs," *BMC Surgery* 14 (2014): 96, https://doi.org/10.1186/1471 -2482-14-96.

4. "Standardizing Emergency Department to Operating Room Handoff for Class A Surgery," University of Iowa Health Care, December 2, 2015, https://medcom.uiowa.edu/theloop/quest-newsletter/standardizing -emergency-department-to-operating-room-handoff-for-class-a-surgery.

5. J. Birkinshaw, "What to Expect from Agile," *MIT Sloan Management Review* 59, no. 2 (winter 2018): 39–42.

6. K. R. McFarland, "Should You Build Strategy Like You Build Software?" *MIT Sloan Management Review* 49, no. 3 (spring 2008): 68–74.

7. M. Sliger, "Agile Project Management with Scrum" (paper presented at PMI Global Congress 2011—North America, Dallas, Texas, October 22, 2011), https://www.pmi.org/learning/library/agile-project-management -scrum-6269.

8. B. Sanner and J. S. Bunderson, "The Truth about Hierarchy," *MIT Sloan Management Review* 59, no. 2 (winter 2018): 49–52.

Chapter 12

1. P. Cappelli and A. Tavis, "The Performance Management Revolution," *Harvard Business Review* 94 (October 2016): 58–67.

2. J. Boudreau and S. Rice, "Bright, Shiny Objects and the Future of HR," *Harvard Business Review* 93 (July–August 2015): 72–78.

3. S. E. Moss and J. I. Sanchez, "Are Your Employees Avoiding You? Managerial Strategies for Closing the Feedback Gap," *Academy of Management Journal* 18, no. 1 (2004): 32–44.

4. A. Rasheed, S. U. R. Khan, M. F. Rasheed, and Y. Munir, "The Impact of Feedback Orientation and the Effect of Satisfaction with Feedback on In-Role Job Performance," *Human Resource Development Quarterly* 26, no. 1 (2015): 31–51.

5. M. M. Lombardo and R. W. Eichinger, *The Leadership Machine* (Minneapolis, MN: Lominger International, 2007), 127.

6. P. G. Audia, E. A. Locke, and K. G. Smith, "The Paradox of Success: An Archival and a Laboratory Study of Strategic Persistence Following Radical Environmental Change," *Academy of Management Journal* 43, no. 5 (2000): 837–853.

7. D. Van-Dijk and A. N. Kluger, "Feedback Sign Effect on Motivation: Is It Moderated by Regulatory Focus?" *Applied Psychology* 53, no. 1 (2004): 113–135.

8. M. London and J. W. Smither, "Feedback Orientation, Feedback Culture, and the Longitudinal Performance Management Process," *Human Resource Management Review* 12, no. 1 (2002): 81–100.

9. C. F. Lam, D. S. DeRue, E. P. Karam, and J. R. Hollenbeck, "The Impact of Feedback Frequency on Learning and Task Performance: Challenging the 'More Is Better' Assumption," *Organizational Behavior and Human Decision Processes* 116, no. 2 (2011): 217–228.

10. J. Wiles, "The Real Impact of Removing Performance Ratings on Employee Performance," Gartner, May 12, 2016, https://www.gartner .com/smarterwithgartner/corporate-hr-removing-performance-ratings -is-unlikely-to-improve-performance/.

11. L. Goler, J. Gale, and A. Grant, "Let's Not Kill Performance Evaluations Yet," *Harvard Business Review* 94, no. 11 (November 2016): 90–94.

12. A. J. Kinicki, G. E. Prussia, B. J. Wu, and F. M. McKee-Ryan, "A Covariance Structure Analysis of Employees' Response to Performance Feedback," *Journal of Applied Psychology* 89, no. 6 (2004): 1057–1069.

13. A. N. Kluger and A. DeNisi, "The Effects of Feedback Interventions on Performance: A Historical Review, a Meta-Analysis, and a Preliminary Feedback Intervention Theory," *Psychological Bulletin* 119, no. 2 (1996): 254–284.

14. H. Song, A. L. Tucker, K. L. Murrell, and D. R. Vinson, "Closing the Productivity Gap: Improving Worker Productivity Through Public Relative Performance Feedback and Validation of Best Practices," *Management Science* 64, no. 6 (December 2016), https://doi.org/10.1287/mnsc .2017.2745.

15. N. Kinley and S. Ben-Hur, "The Missing Piece in Employee Development," *MIT Sloan Management Review* 58, no. 4 (summer 2017): 89–90.

16. Wiles, "The Real Impact of Removing Performance Ratings."

17. Lombardo and Eichinger, *The Leadership Machine*, 133.

18. Kinicki, Prussia, Wu, and McKee-Ryan, "Covariance Structure Analysis," 89.

19. R. D. Pritchard, S. S. Youngcourt, J. R. Philo, D. McMonagle, and J. H. David, "The Use of Priority Information in Performance Feedback," *Human Performance* 20, no. 1 (2007): 61–83.

20. C. M. Kuhnen and A. Tymula, "Feedback, Self-Esteem, and Performance in Organizations," *Management Science* 58, no. 1 (2012): 94–113.

21. S. Ertac, "Does Self-Relevance Affect Information Processing? Experimental Evidence on the Response to Performance and Non-Performance Feedback," *Journal of Economic Behavior & Organization* 80, no. 3 (2011): 532–545.

22. D. VandeWalle, W. L. Cron, and J. W. Slocum Jr., "The Role of Goal Orientation Following Performance Feedback," *Journal of Applied Psychology* 86, no. 4 (2001): 629–640.

23. Rasheed, Khan, Rasheed, and Munir, "The Impact of Feedback Orientation."

24. R. Hogan and R. B. Kaiser, "What We Know About Leadership," *Review of General Psychology* 9, no. 2 (2005): 169–180.

25. R. B. Kaiser, ed., *The Perils of Accentuating the Positive* (Tulsa, OK: Hogan Press, 2009).

26. K. L. Sommer and M. Kulkarni, "Does Constructive Performance Feedback Improve Citizenship Intentions and Job Satisfaction? The Roles of Perceived Opportunities for Advancement, Respect, and Mood," *Human Resource Development Quarterly* 23, no. 2 (2012): 177–201.

27. J. S. Goodman, R. E. Wood, and Z. Chen, "Feedback Specificity, Information Processing, and Transfer of Training," *Organizational Behavior and Human Decision Processes* 115, no. 2 (2011): 253–267.

28. C. F. Lam, D. S. DeRue, E. P. Karam, and J. R. Hollenbeck, "The Impact of Feedback Frequency on Learning and Task Performance," *Organizational Behavior and Human Decision Processes* 116, no. 2 (2011): 217–228.

29. B. Kuvaas, R. Buch, and A. Dysvik, "Constructive Supervisor Feedback Is Not Sufficient: Immediacy and Frequency Is Essential," *Human Resource Management* 56, no 3 (2017): 519–531.

30. A. S. Tsui and P. Ohlott, "Multiple Assessment of Managerial Effectiveness: Interrater Agreement and Consensus in Effectiveness Models," *Personnel Psychology* 41, no. 4 (1988): 779–803.

31. C. O. Longenecker, H. P. Sims Jr., and D. A. Gioia, "Behind the Mask: The Politics of Employee Appraisal," *Academy of Management Journal* 1, no. 3 (1987): 183–193.

32. A. M. Lane and S. Gorbatov, "Fair Talk: Moving Beyond the Conversation in Search of Increased and Better Feedback," *Performance Improvement* 56, no. 10 (2017) 6–14.

33. J. Welch, "Jack Welch: 'Rank-and-Yank'? That's Not How It's Done," *Wall Street Journal*, November 14, 2013, https://www.wsj.com/articles/8216rankandyank8217-that8217s-not-how-it8217s-done-1384473281.

34. S. E. Moss and J. I. Sanchez, "Are Your Employees Avoiding You?" *Academy of Management Journal* 18, no. 1 (February 2004): 32–44.

35. Lombardo and Eichinger, *The Leadership Machine*, 78.

Chapter 14

1. S. Subramanian, "Google Took Its 20% Back, But Other Companies Are Making Employee Side Projects Work for Them," Fast Company, August 19, 2013, https://www.fastcompany.com/3015963/google-took-its-20-back-but-other-companies-are-making-employee-side-projects-work-for-them.

2. N. Anderson, K. Potocnik, and J. Zhou, "Innovation and Creativity in Organizations: A State-of-the-Science Review, Prospective Commentary, and Guiding Framework," Journal of Management 40, no. 5 (July 2014): 1297–1333.

3. E. L. Deci and R. M. Ryan, *Intrinsic Motivation and Self-Determination in Human Behavior* (New York: Plenum Press, 1985); and M. Gagné and E. L. Deci, "Self-Determination Theory and Work Motivation," *Journal of Organizational Behavior* 26, no. 4 (June 2005): 331–362.

4. Researchers more often use the term "prosocial" to describe this kind of motivation, but "social" conveys an appropriate intuitive meaning. See A. M. Grant and J. W. Berry, "The Necessity of Others Is the Mother of Invention: Intrinsic and Prosocial Motivations, Perspective Taking, and Creativity," *Academy of Management Journal* 54, no. 1 (February 2011): 73–96.

5. R. D. Austin and C. Stormer, "Miles Davis: Kind of Blue," Harvard Business School case no. 609–050 (Boston: Harvard Business School Publishing, 2008).

Chapter 16

1. J. M. Kouzes and B. Z. Posner, *Credibility: How Leaders Gain and Lose It, Why People Demand It* (San Francisco, CA: Jossey-Bass, 2011); and J. M. Kouzes and B. Z. Posner, *The Leadership Challenge: How to Make*

Extraordinary Things Happen in Organizations, 6th ed. (Hoboken, NJ: John Wiley & Sons, 2017).

2. Political leaders lose their credibility as well, and some (such as Brazilian president Dilma Rousseff and South Korean president Park Geun-hye) have been removed from office and even imprisoned.

3. See T.-Y. Kim, T. S. Bateman, B. Gilbreath, and L. M. Andersson, "Top Management Credibility and Employee Cynicism: A Comprehensive Model," *Human Relations* 62, no. 1 (August 2009): 1435–1458; and P. H. Kim, K. Dirks, C. D. Cooper, and D. L. Ferrin, "When More Blame Is Better Than Less: The Implications of Internal vs. External Attributions for the Repair of Trust after a Competence- vs. Integrity-Based Trust Violation," *Organizational Behavior and Human Decision Processes* 99, no. 1 (2006): 49–65.

4. J. M. Kouzes and B. Z. Posner, "Leading in Cynical Times," *Journal of Management Inquiry* 14, no. 4 (2005): 357–364; D. G. Leathers, *Successful Nonverbal Communication: Principles and Applications* (New York: Macmillan, 1992); R. C. Mayer, J. H. Davis, and F. D. Schoorman, "An Integrative Model of Organizational Trust," *Academy of Management Review* 20, no. 3 (1995): 709–734; and T. L. Simons, "Behavioral Integrity: The Perceived Alignment between Managers' Words and Deeds as a Research Focus," *Organization Science* 13, no. 1 (2002): 18–35.

5. K. T. Dirks, "Three Fundamental Questions Regarding Trust in Leaders," in *Handbook of Trust Research,* ed. R. Bachmann and A. Zaheer (Cheltenham, U.K.: Edward Elgar, 2006), 15–28; and P. H. Kim, D. L. Ferrin, C. D. Cooper, and K. T. Dirks, "Removing the Shadow of Suspicion: The Effects of Apology versus Denial for Repairing Competence- versus Integrity-Based Trust Violations," *Journal of Applied Psychology* 89, no. 1 (2004): 104–118.

6. B. M. Bass, *Bass & Stogdill's Handbook of Leadership: Theory, Research, and Managerial Applications,* 3rd ed. (New York: Free Press, 1990).

7. Kouzes and Posner, *Credibility.*

8. "Patagonia's Mission Statement," n.d., https://www.patagonia.com/company-info.html.

9. Reuters, "Buffett Says Wells Fargo Was 'Slow to Fix' Sales Scandal," May 6, 2017, https://www.newsweek.com/warren-buffett-wells-fargo-berkshire-hathaway-berkshire-wells-fargo-scandal-595892.

10. "BP Boss Tony Hayward's Gaffes," *BBC*, June 20, 2010, https://www
.bbc.com/news/10360084.

11. T.-Y. Kim et al., "Top Management Credibility"; and P. H. Kim et al.,
"Removing the Shadow of Suspicion."

12. Dirks, "Three Fundamental Questions"; and P. H. Kim et al.,
"Removing the Shadow of Suspicion."

13. P. H. Kim et al., "When More Blame Is Better Than Less."

Chapter 17

1. R. Parloff, "How VW Paid $25 Billion for 'Dieselgate'—and Got Off
Easy," *Fortune*, February 6, 2018, http://fortune.com/2018/02/06/volks
wagen-vw-emissions-scandal-penalties/.

2. "The Wells Fargo Fake Accounts Scandal Just Got a Lot Worse," *Fortune*, August 31, 2017, http://fortune.com/2017/08/31/wells-fargo
-increases-fake-account-estimate/.

3. "Samsung Heir Lee Jae-yong Jailed for Corruption," *BBC*, August 25,
2017, https://www.bbc.com/news/business-41033568.

4. T. Haugh, "The Trouble with Corporate Compliance Programs," *MIT
Sloan Management Review* 59, no. 1 (fall 2017): 55–62.

5. D. R. May, Y. K. Chang, and R. Shao, "Does Ethical Membership
Matter? Moral Identification and Its Organizational Implications," *Journal of Applied Psychology* 100, no. 3 (2015): 681–694; and O. Demirtas
and A. A. Akdogan, "The Effect of Ethical Leadership Behaviour on Ethical Climate, Turnover Intention, and Affective Commitment," *Journal of
Business Ethics* 130 (2015): 59–67.

6. T. W. Ng and D. C. Feldman, "Ethical Leadership: Meta-Analytic Evidence of Criterion-Related and Incremental Validity," *Journal of Applied
Psychology* 100, no. 3 (2015): 948–965.

7. C. Bailey, A. Shantz, P. Brione, R. Yarlagadda, and K. Zheltoukhova,
Purposeful Leadership: What Is It, What Causes It, and Does It Matter?
(London: Chartered Institute of Personnel and Development, June
2017).

8. We draw here on the ideas of W. Mischel, *Personality and Assessment*
(New York: Wiley, 1968).

9. A. L. Kristof-Brown, R. D. Zimmerman, and E. C. Johnson, "Consequences of Individuals' Fit at Work: A Meta-Analysis of Person–Job,

Person–Organization, Person–Group, and Person–Supervisor Fit," *Personnel Psychology* 58, no. 2 (June 2005): 281–342.

10. M. Motyl, R. Iyer, S. Oishi, S. Trawalter, and B. A. Nosek, "How Ideological Migration Geographically Segregates Groups," *Journal of Experimental Social Psychology* 51 (2014): 1–14.

11. C. Frisch and M. Huppenbauer, "New Insights into Ethical Leadership: A Qualitative Investigation of the Experiences of Executive Ethical Leaders," *Journal of Business Ethics* 123, no. 1 (2014): 23–43.

12. J. Pucic, "Do as I Say (and Do): Ethical Leadership through the Eyes of Lower Ranks," *Journal of Business Ethics* 129, no. 3 (2014): 655–671.

13. J. McGregor, "More CEOs Are Getting Forced Out for Ethics Violations," *Washington Post*, May 15, 2017, https://www.washingtonpost .com/news/on-leadership/wp/2017/05/15/more-ceos-are-getting-forced -out-for-ethics-violations/.

14. Parloff, "How VW Paid $25 Billion for 'Dieselgate.'"

15. S. A. Eisenbeiss, D. van Knippenberg, and C. M. Fahrbach, "Doing Well by Doing Good? Analyzing the Relationship between CEO Ethical Leadership and Firm Performance," *Journal of Business Ethics* 128 (2014): 635–651.

16. Haugh, "The Trouble with Corporate Compliance Programs."

17. J. R. Detert and E. R. Burris, "Can Your Employees Really Speak Freely?" *Harvard Business Review* 94, no. 1 (January–February 2016): 80–87.

18. Parloff, "How VW Paid $25 Billion for 'Dieselgate.'"

19. Ibid.

Index